SUMMER FRUIT-HOUSE.

FROM PLANS BY

JOHN C. SCHOOLEY,

CINCINNATI, OHIO.

(PROCESS PATENTED MARCH 13, 1855.)

Fig. 1. Scale, ¾ in. to 10 feet.

FIG. 1.—View of an ICE and FRUIT-HOUSE, under one roof, built entirely above ground, with dormer-windows in roof, three of which are counterfeit; the other one is a door to admit ice, with lattice transom, made so as to admit outside atmosphere on to the ice.

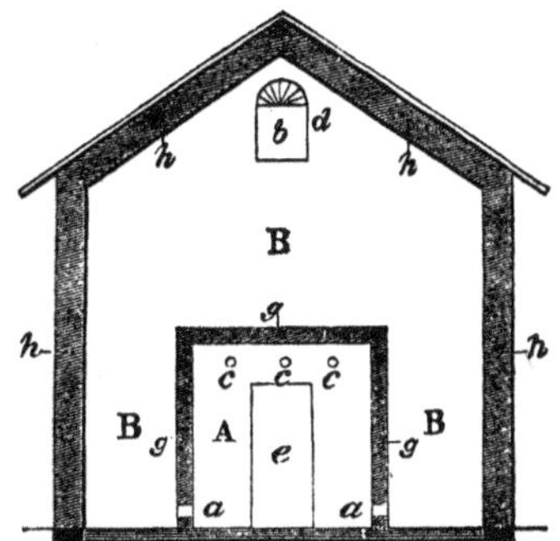

Fig. 2.

Fig. 2.—A perpendicular sectional view of same.

B. Ice-room, over and at the side of the fruit-room.

A. Fruit-room. e. Inner door to same. c, c, c. Small round openings for the egress of the air from fruit-room. a, a. Flues from ice-house to fruit-room. g. Partition between ice and fruit rooms, 12 inches thick (would be better to be made 18 inches thick). b. Door in dormer-window, to admit ice. d. Lattice-work opening to admit air on to ice.

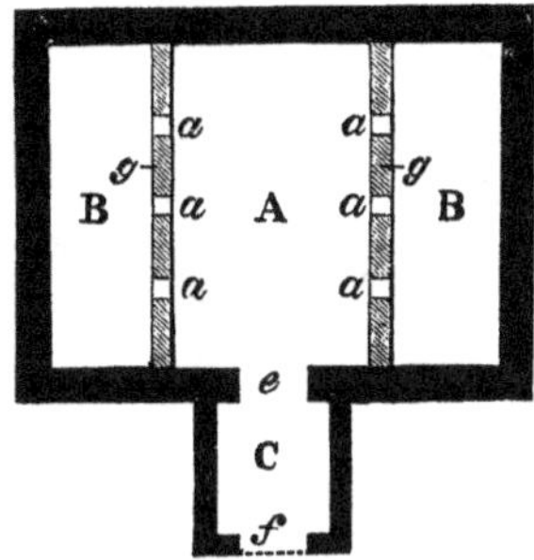

Fig. 3.

Fig. 3.—Horizontal, sectional view of ground-plan of the structure.

A. Fruit-room.

B, B. Ice-house.

C. Vestibule, or protection-porch, to keep the outside atmosphere from penetrating into fruit-room. f. Outside door of same, well insulated. e. Inside, and entrance to fruit-room, well insulated. a, a. Six openings for the admission of cold dry air from ice, each opening one foot square, with slides to open and close at will. g. Partition, well insulated.

This house can be built with either brick or wood; the latter is preferable. The fruit and ice house can be constructed within another building.

N. B. Particular care should be taken in all cases to have the bottoms of these houses thoroughly drained, and well insulated from the natural heat of the earth. This is absolutely necessary, as heat affects the earth to the depth of forty feet.

SUMMER, FRUIT, ICE, AND ARTIFICIAL SPRING HOUSE, FROM PLANS BY JOHN C. SCHOOLEY, CINCINNATI, OHIO.

(Process Patented March 13, 1855.)

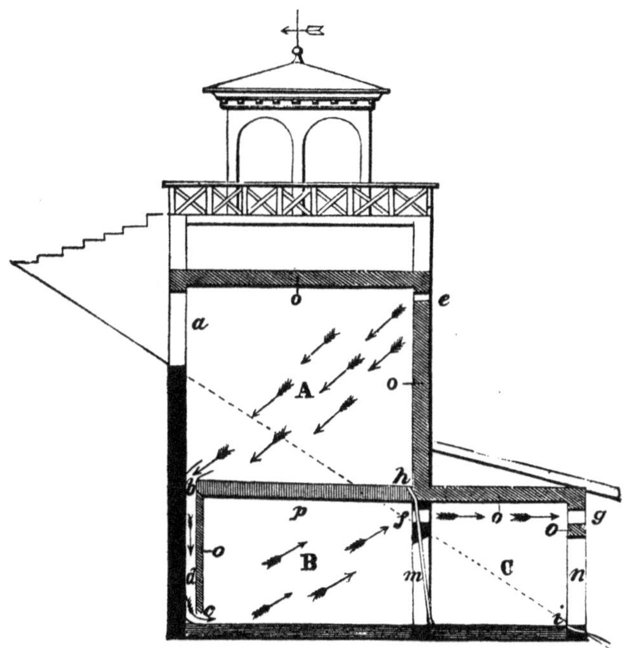

Fig. 1. Scale, ¾ in. to 10 feet.

FIG. 1.—Represents a sectional view of SUMMER, ICE, FRUIT, and ARTIFICIAL SPRING HOUSE, located on side hill.

A. Ice-house, twenty by twenty-five feet (outside). a. Door into ice-house, to admit ice, two and a half feet long and one foot wide, made with a slide to close and open at will; if it is more convenient, it can be made in the door a.

B. Fruit-room, twenty by twenty-five feet (outside).

C. Artificial Spring-House. p. Inclined floor of ice-house, well insulated, eighteen inches thick, well caulked or covered with zinc. b. Mouth of descending flue, twelve inches high, and to extend the entire width of ice-room, as in fig. 2. d. Descending flue. c. Mouth of same, in fruit-room. f. Opening for the escape of air, from fruit-room into spring-house (one foot high, and one foot six inches long). g. Outlet opening (one foot by two feet). m. Door from spring-house into fruit-room. n. Door from outside into spring-house. h. Mouth of lead pipe, to lead off the ice-meltings down into spring-house. i. Escape-channel for the water to run out and down side hill. o. Partition, to be made eighteen inches thick, and to be filled with dry saw-dust, or tan. The light-colored partitions are to be made of wood, well lined with close boards, or flooring. The dark part is a stone wall, being against and in the earth. The stone wall should be lined with rough, dry boards, on the inside. The fruit-room, B, should be either lathed and plastered, or lined with dry pine flooring.

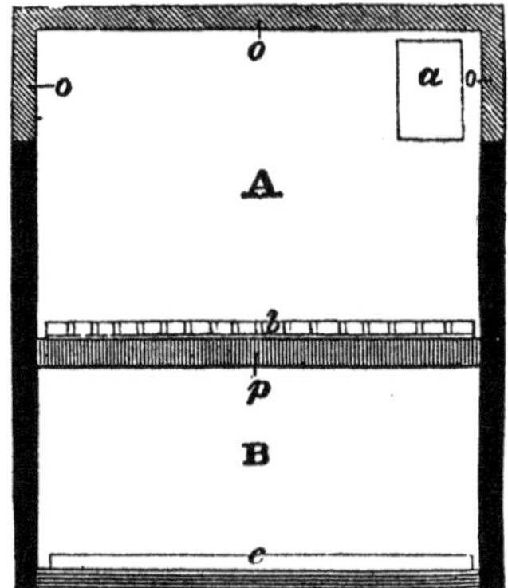

Fig. 2.

FIG. 2.—Is a perpendicular sectional view of the side next to the hill; a, ice-house door, corresponding with a in Fig. 1. b, mouth of descending flue, showing entire length of same, with bars over it, to prevent the ice from falling down the flue. c, mouth of same in fruit-house, with sliding valve to regulate the draft. See Fig. 1.

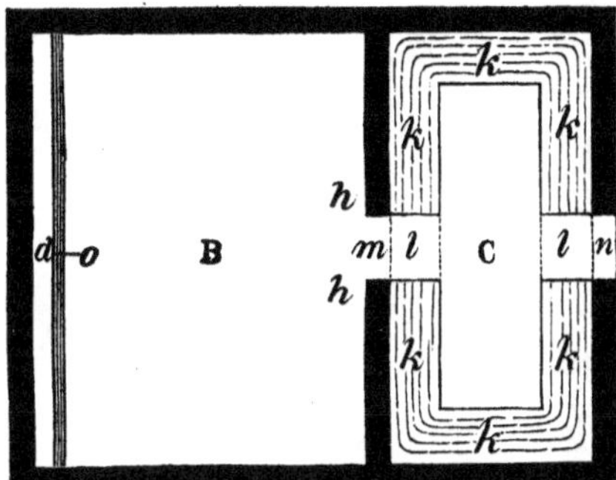

FIG. 3.

FIG. 3.—Is a horizontal sectional view of the ground plan. h, h is place of entrance of descending pipe, or pipes for the ice-meltings. l, l is temporary floor over channel. k, is a channel around the bottom, and next to the sides of the spring-house, two feet wide, and one foot deep, bricked or cemented up, to contain always a sufficient quantity of the cold ice water from meltings, which can escape out under door n, as represented by letter i, in fig. 1. This escape opening must be so constructed as to permit a certain depth of water to always remain in channel k; yet continually escaping, thus having a stream of fresh ice water flowing through the spring-house at all times. Crocks of butter, and pans of milk, can be placed in this channel of running water, where they will be kept fresh and cold for almost any length of time.

The style of the building may conform to the taste of individuals. It may be ornamented like the design in fig. 1, with Summer-house on top, or be attached to other structures; or in place of artificial spring-house, there may be an ante-room, or porch, to keep outside atmosphere from penetrating into Fruit-room.

HOOPER'S

WESTERN FRUIT BOOK.

The following opinions are from gentlemen well known as eminent Horticulturists, who examined the work in manuscript:

SPRING GARDEN, CINCINNATI,
November 12th, 1856.

E. J. HOOPER, Esq.,

My Dear Sir—I have given your manuscript Catalogue of Fruits a hasty examination, and regret that my time does not admit of a more deliberate and critical one. The arrangement is admirable, and can not fail to be immensely valuable to cultivators, as well as to those who are making selections of Fruit Trees for planting. The great confusion which exists in the nomenclature is exceedingly embarrassing. This your arrangement removes, and makes the subject easily understood. Your descriptions, though brief, are pointed, and really all that is needed, to guide the uninformed in making judicious selections. The number of varieties that have been brought to notice within the last few years, is so great, and the want of time to test them in our soil and climate so short, that such a work as yours will require great care and labor; and even then will of necessity require several editions, corrections, and additions, to approach anything like perfection.

With my best wishes for your success in the noble undertaking, I remain, very respectfully, yours,

A. H. ERNST.

CINCINNATI, October 17, 1856.

E. J. HOOPER, Esq.,

Dear Sir — I have examined your Catalogue of Fruits carefully; and, so far as I am able to judge, I believe the nomenclature to be accurate, and most of the synonyms correctly given. I allude to the Fruits generally known and tested — for new seedling varieties, especially of Apples, multiply on us so fast, that it is hard to classify them, and give them a proper place in our catalogues.

I agree with Dr. Warder, Mr. Ernst, and Professor Cary, in their notes on your Fruits, and think them accurate and just. The points of difference, if any, are so few that I will not designate them.

That your Catalogue will be valuable to Fruit Growers and Horticulturists, there can be no doubt; and I am much pleased to find, that you are about to supply, in a compact form, so brief and ready a mode of reference to the Fruits cultivated in the West. It is much wanted.

Very respectfully,

R. BUCHANAN.

LATONIA SPRINGS, January 22, 1857.

E. J. HOOPER,

Dear Sir — I have carefully looked over your manuscript of Apples, in your contemplated work on Pomology, and, with a few alterations which I have taken the liberty, by your permission, to suggest, believe that your book will be of great utility to those engaged in the pursuit of Pomology. I like the plan of your work, and think that it will be well adapted to this vicinity, and to Western localities, generally.

The descriptions of Apples are generally concise, and well drawn, and give the characters of the fruit very clearly — of all the most valuable kinds, especially. Many new varieties are described of which I have little knowledge, and therefore can not judge of their accuracy.

I think a work like yours, descriptive of the Fruits of this section of country, with the soils best adapted to their growth, very much needed.

Yours, respectfully,

S. MOSHER.

Middleton, Wallace & Co. Lithrs. Cin. O.

JOHN P. FOOTE. A. H. ERNST. F. PENTLAND. E. J. HOOPER. S. S. JACKSON. J. SAYER. Dr. J. A. WARDER.

GABRIEL SLEATH. GEO. GRAHAM. R. BUCHANAN. N. LONGWORTH.

HOOPER'S

WESTERN FRUIT BOOK:

A COMPENDIOUS

COLLECTION OF FACTS,

FROM THE

NOTES AND EXPERIENCE

OF

SUCCESSFUL FRUIT CULTURISTS,

ARRANGED FOR PRACTICAL USE

IN

THE ORCHARD AND GARDEN.

"Under these general laws, each variety of fruit requires a particular treatment, and should be nurtured with a wise reference to its peculiarities and habits."
Hon. MARSHALL P. WILDER, *Pres't Mass. Hort. Society.*

BY E. J. HOOPER,

MEMBER OF THE CINCINNATI HORTICULTURAL SOCIETY, AND FORMERLY EDITOR OF THE "WESTERN FARMER AND GARDENER."

CINCINNATI:
MOORE, WILSTACH, KEYS & CO.,
25 WEST FOURTH STREET.
1857

Stereotyped and Printed by
MOORE, WILSTACH, KEYS & CO.,
CINCINNATI, O.

TO

DR. JOHN A. WARDER,

PRESIDENT OF THE CINCINNATI HORTICULTURAL SOCIETY,

This Volume is Dedicated

BY HIS FRIEND,

THE AUTHOR.

INDEX TO FRUITS.

NOTE EXPLANATORY.

The reader will observe that we have classified the fruits in this work as follows:

The best in quality, flavor, etc., or No. 1, in CAPITALS.
Second best " " " " 2, in SMALL CAPS.
Third " " " " " 3, in *italics.*

PREFACE.

It is the first step in science to know what is known. What is new, and dependent on experience and original observation, will then come easier and more certainly. It is an economy of time and labor, in any investigator, to ascertain well what has been done before him, in any field of experiment. Much time is often irrecoverably wasted in blundering over proposed experiments, and supposed novelties, that have long before been thoroughly examined and definitely settled. In no branch of practical science, are these maxims more true than in regard to the cultivation of fruit; and this because there are so many claims upon the cultivator's attention; so many drafts upon his credulous inexperience; so many contradictory statements resulting from superficial investigations; so many delusory appearances; so much pretension and self-serving; so much that rests upon inadequate and interested evidence. There is, in a word, so much to confuse, mislead and deceive, that he who shall present to the fruit-grower, a key to these conflicting claims and representations, giving, in words of truth and soberness, a just and concise statement of what may be relied on as fact, in regard to the value and names of such fruits as are really and honestly known to be worthy of acceptation and confidence,—that man will have done a good work, and should be welcomed of all men as a benefactor, in a field where ignorance is attended with innumerable mischievous consequences, and where doubt is about as fatal as ignorance.

There have been several praiseworthy laborers in this inviting field, and all with more or less fault and excellence, more or less accuracy and error—the result, perhaps, of too much haste in compilation, and too great confidence in mistaken and interested testimony. None have seemed exactly to fill the purpose desired; and the want of a new and more competent work, in the shape of a concise and reliable hand-book, was very generally experienced, and widely and repeatedly expressed. A work was needed less voluminous and less diffusive, based upon the authentic experience of actual cultivators; upon well-purged lists of fully proved and living trees, whose fruit had been properly tested and characterized, and of whose identification there was, finally, no question. A book unincumbered with useless descriptions of worthless varieties, and unneeded directions for planting and cultivation, and free from all guess-work, and all unverified statements, and confusing and half-recognized synonyms. A descriptive and concise list, in fact, of such actual fruits, as are well established, and clearly identified, with their most generally accepted names, and their most marked and unmistakable characteristics.

And this was the plan designed and undertaken by the practical and practiced author of the present work. Himself a fruit-grower of diversified experience, and having been in correspondence for years with some of the most prominent and successful cultivators in the United States, especially with those whose experience has chiefly related to the peculiarities and requisitions of the Middle and North-Western States, and feeling, in his own practice, the want of such a ready guide as the one contemplated in his plan, he came to the work, prepared with his own accumulated observations, the advice and suggestions of other competent growers, and the advantage of the several larger, but differently designed, works that had preceded his own. The errors of these last he was to correct, and their faults he was to avoid. The task was by no means an easy one, and would not have been undertaken but for the steady encouragement of many warm friends of pomological science, and the aid of many efficient cultivators; in whose knowledge and candor he had steady reason to confide, and for whose kindness he desires, here, to acknowledge his repeated indebtedness.

That the work is faultless, is not claimed. That it will be found convenient, thorough and accurate, and just adapted to daily field use, by the Western grower of both large and small fruits, whether professional or amateur, is fully believed by the accomplished author's friend and associate.

J. W. W.

CINCINNATI, February 15, 1857.

HOOPER'S

WESTERN FRUIT BOOK.

APPLES.

ALEXANDER, synonymous with, or called by some *Emperor Alexander*, *Russian Emperor*, and *Aporta*; color, streaked with bright red on greenish yellow; form, sometimes slightly conical; base, somewhat flattened; size, 1; use, chiefly kitchen; quality, 2; season, August to December.

REMARKS.—Very large and handsome. A moderate bearer. Rather coarse, but very beautiful in color, shape, and has a fine bloom. Succeeds well in the Western States. A good grower. "Large and beautiful."—*Transactions Ohio Pomological Society.*

ALFRISTON, *Newtown Pippin of some in England*; color, greenish yellow; form, roundish; size, 1; use, kitchen; quality, nearly 3; season, September to December.

REMARKS.—Foreign. No Similarity to Newtown Pippin in America, as thought by some in England.

AMERICAN GOLDEN RUSSET, or called by some *Bullock's Pippin, Sheepnose, Golden Russet,* and *Little Pearmain*, sometimes, erroneously, *Fall Winesap*; color, generally rich golden yellow, overspread with soft russet, and in the sun a little red; form, roundish ovate, tapering toward the eye; size, 3; use, table; quality, 1; season, December to February.

REMARKS.—This delicious apple succeeds well in the locality of Cincinnati, and in Indiana, in rich soils. "First-rate and handsome."—*Trans. Ohio Pom. Society.*

AMERICAN MAMMOTH. Synonymes, *Ox Apple, New York Gloria Mundi, Baltimore Pippin,* and *Gloria Mundi,* which last title see also.

AMERICAN SUMMER PEARMAIN, or *Watkin's Early and American Pearmain.* Color, greenish yellow, with a little red; form, pearmain, or roundish oblong; size, 2; use, table; quality, 1; season, August and September.

REMARKS.—This deserves to be called Summer King, compared with the Summer Queen. An abundant bearer. Different from the English Pearmain. It is good in nearly all parts of the country. "Highly approved."—*Trans. Ohio Pom. Society.*

American Pippin, or Grindstone. Color, greenish red, with red streaks; form, round and flattish; size, 2; use, kitchen and table; quality, 3; season, January to July.

REMARKS.—There are many of this name, which is rather vague, though expressive. It is a good bearer, and a very great keeper, but almost totally unworthy; uneatable at any time.

APPLE BUTTER, or *Sweet Bellflower of some,* and *Molasses of others.* Color, yellow; form, roundish, conical; size, 1; use, table and kitchen; quality, 1.

REMARKS.—There are two apples cultivated in Southern Ohio under this name. Mr. A. H. Ernst values them highly. Dr. Warder, one of our best pomologists, considers them tender and very good. Exhibited by F. G. Cary, August, 1855. Mr. Stikes, of Dayton, ranks one of them also high.

Api. See Lady Apple.

Ashland. Color, dull greenish yellow; form, roundish; size, 2; use, table; quality, 2; season, October to January, often longer.

Remarks.—A sound, good keeper. Considered of a good flavor by most. Exhibited before the Cincinnati Horticultural Society by R. Buchanan. "A sound, sweet-tasted apple, of medium size."—*Fruit Committee.*

Ashmore, sometimes erroneously called *Fall Wine.* Color, bright, clear red; form, regular roundish, flattened; size, 2 to 1; use, table; quality, 2; season, September and October.

Remarks. — White, crisp, tender, juicy, sub-acid, sprightly. Not of high character, but tender. Exhibited before Cincinnati Horticultural Society, by A. H. Ernst, August 21, 1855.

Astrachan Red. Color, crimson red; form, roundish; size, 2; use, table, or dessert; quality, 2; season, July to August.

Remarks.—A very handsome, rather acid dessert fruit, with a bloom on it similar to a plum. It cooks well, and is productive. A good fruit for market. Exhibited by R. Buchanan before the Horticultural Society, July, 1855. Fruit Committee decided it, "Beautiful and very good."

Autumn Pearmain, or *English Summer Pearmain*, *Royal Pearmain*, *Sigler's Red*, etc. Color, brownish yellow, green and red; form, oblong; size, 2; use, table; quality, 2; season, August and September.

REMARKS.—This is our Autumn Seek-no-further. The tree is a slow grower. Branches slender. Flesh crisp, firm, and a little dry.

AUTUMNAL SWAAR. Color, pale green, sometimes a slight red in sun; form, roundish; size, 2; use, table; quality, 1; season, October.

REMARKS.—This is regarded by Dr. Mosher, of Latonia Springs, one of our best eating apples. The grain is very fine; juicy, tender, sprightly, and sub-acid in flesh. If two-thirds of this apple rots, the remainder retains the fine juice and flavor.

BAILEY SPICE. Color, yellow; form, roundish; size, 2; use, table; quality, 2; season, August and September.

REMARKS.—Fruit always fair at Rochester, New York; moderate growth; flesh, sprightly, spicy.

BAILEY SWEET, or *Patterson Sweet, Edgerly Sweet, etc.* Color, clear yellowish red, with russet patches; form, round ovate, flattened; size 1; use, table; quality, 2.

REMARKS.—Grown in Northern Illinois, where, as with us, it is beautiful, delicate, sweet, juicy, and rich.

BALDWIN. Color, red and orange, brilliant; form, roundish oblate; size, 1 to 2; use, table; quality, 1 (where it succeeds); season, October to March.

REMARKS.—Of the Æsopus Spitzenburg family. It is the great Boston Apple. It is rather subject to rot in Ohio and Kentucky. "Good at the North; subject to bitter rot at the South."—*Trans. Ohio Pom. Soc.* In this locality (Cincinnati), not generally successful as a keeper, though sometimes seen very fine in January; generally an early winter apple, subject to worms and rot, and never to be compared to Æsopus Spitzenburg for flavor. The tree bears early and well; a vigorous grower; tolerably upright; spreads when older.

Bartlett, synonymous with *Priestley, or Bullet.* Color, bright red on yellow; form, oblong; size, 2; use, table; quality, 3; season, January to June.

REMARKS.—"Long keeper."—*Trans. Ohio Pom. Society.* Rather an inferior fruit.

Beauty of the West, Red Bellflower of some, and Ohio Nonpareil, Wells, etc. Color, green and yellow; form, roundish, flat at base; size, 1 to 2; use, kitchen; quality, 2 to 3; season, September and October.

REMARKS.—It is sweet, and keeps some time. It is handsome but rather poor. It is not known as Red Bellflower at Cincinnati, or in the south of Ohio. A remarkably fine grower.

Beauty of Kent. Color, striped with purple, red, greenish yellow; form, roundish; size, 1; quality, 3; season, August and September.

REMARKS.—Rivals the Alexander in size, but not so good in flavor. The tree is vigorous and productive. Exhibited at Cincinnati Horticultural Society, by A. H. Ernst, August 21, 1855.

BELLFLOWER, WHITE, or *Detroit, Ohio Favorite, Ortley* by Lindley, etc. Color, pale yellowish white; form, oblong oval, or roundish conical; size, 1 to 2; use, table; quality, nearly 1; season, December to April.

REMARKS.—Large on rich soils; core, open. "Excellent on most strong soils."—*Trans. Ohio Pom. Soc.* It is often affected in South Ohio with the bitter rot, but where this is not the case it is one of the best apples in South Ohio and Indiana. There is now a superbly fine seedling of this variety, raised by Mr. Davis, in Southern Indiana. This seedling has the advantage of not being hollow in the core, like the parent fruit, the White Bellflower. It is named *Davis' Seedling.*

BELLFLOWER, YELLOW. Color, pale yellow, with a blush next the sun; form, oblong; size, 1; use, table; quality, 1; season, October to March.

REMARKS.—A beautiful, peculiarly good flavored, and well known apple in the markets of Cincinnati and the West. The wood is slender, and like the Newark, or French Pippin, bears its fruit on the ends of the limbs. It is desirable to graft it above ground. The blossoms are very beautiful. It is a superior variety, but not a great bearer. It blooms early, and before the leaves expand, that is, on long stems, and therefore liable to be frosted. It bears well; the fruit drops a little, but all are good for cooking at all times. It becomes of less value as it is grown toward the north of us.

BELMONT, or *Gate.* Color, rich, light yellow, with a glossy surface, and a bronzy blush toward the sun; form, roundish oblong, but irregular; size, 2; use, dessert; quality (Cincinnati), 1 to 2; season, October to February.

REMARKS.—This is a great and favorite apple in Northern Ohio, but succeeds indifferently in South-Western Ohio. It is not yet fully tested in this locality. Dr. Warder considers it a first-rate apple, if it can be so called without the highest flavor. "Generally approved, especially in Northern Ohio."—*Trans. Ohio Pom. Soc.* The Gate is very fine at Marietta, Ohio. It is showy, and looks quite rich on the stalls.

BENONI. Color, striped red; form, round; size, 2; use, table; quality, 1; season, July.

REMARKS.—Excellent; one of the best early fruits; flesh, yellow; a good bearer; sub-acid, and pleasant. Often exhibited before the Cincinnati Horticultural Society; considered by the Fruit Committee of the Society, "prolific, pleasantly acid, and very agreeable." A strong, upright, good grower. "Handsome, early and good."—*Trans. Ohio Pom. Society.* Comes into bearing early.

Bentley's Sweet. Color, red and yellow striped, or blotched; form, oblong, irregular, flattened at ends; season, January to September.

REMARKS.—"Good as a long keeper."—*Trans. Ohio Pom. Society.*

Bevan. Color, broad red stripes on yellow ground; form, roundish, flattened, slightly conical; size 3; use, table; quality, 3; season, July and August.

REMARKS.—Origin, Ohio. Best for market purposes to carry long distances. But at any rate of but little value, when there are so many so much better. "Of but little value."—*Trans. Ohio Pom. Society.*

Beefsteak.

REMARKS.—"Not approved."—*Trans. Ohio Pom. Society.*

BETTER-THAN-GOOD. Size, 2; use, table; quality, 2.

REMARKS.—Bears too much. A Pennsylvania apple.

Bethlehemite. Color, striped red; form, regular; size, 2; season, October till April.

REMARKS.—"Resembles the Newtown Spitzenburg in appearance."—*Trans. Ohio Pom. Society.* Scarcely worthy of introduction and propagation. A fine grower.

2

Birmingham.

Remarks.—Pronounced by the National Pomological Society "Good."

Black Gilliflower. "Unworthy of culture."—*Trans. Ohio Pom. Society. Black Gilliflower* of Dr. Mosher. Color, very dark red, almost black; form, long, conical; a No. 1 apple; core, hollow, fine flavor, fine perfume; only fault, rather dry, calculated for a more Southern climate, and would be there more juicy.

Black Apple. Color, deep red; form, round; size, 2; use, table; quality, 2; season, August and September.

Remarks.—There are several in the West of this kind. It is a good apple, but it becomes insipid toward Spring. It is sweetish. The tree bears well, and has a round head. Jersey Black is a good fruit among those of this name. A great bearer, and a fair second-rate apple. Stands package and carriage well. In Michigan the Black Apple is known as the Detroit.

Black Detroit. Color, dark blackish crimson; form, roundish, flattened; size, 2; use, table; quality, 2.

Remarks.—Very similar to the above, but is larger, and has more character.

Black Vandevere. Color, very dark red; form, lopsided; size, 2; use, table; quality, 2.

Remarks.—"A good keeper, and second-rate."—*Trans. Ohio Pom. Society.*

Bledsoe.—Color, greenish yellow; form, conical, and rather globular; size, 2; use, table; quality, 2; season, March and April.

Remarks.—Seedling from Kentucky, from Col. Lewis

Sanders, of Gallatin county. "Very like White Pippin." —*A. H. Ernst's MSS.*

Blenheim Orange. Color, yellow and dull red; size, 1; use, table; quality, 2 to 3; season, August and Sept.

Remarks.—Very like Fall Pippin. Exhibited by R. Buchanan, August, 1855.

Blood. Color, dull red; form, round; size, 2; use, table; quality, 2; season December to March.

Blockley Pippin.

Remarks. — Large and productive, of second-rate quality.

Blue Pearmain. Color, purple red; form, round; size, 1; use, table; quality, 2; season, September to January.

Remarks.—That grown by Ellwanger & Barry is not correct. The true is very large, of a dark red bloom. It is a poor bearer. Exhibited before Cincinnati Horticultural Society, July, 1855. Fruit Committee "consider it of much value."

BOHANON. Color, pale green, slight blush near sun; form, round, flattened; size, 2; use, table; quality, 1; season, August.

Remarks.—Brought to Cincinnati Horticultural Society by Lewis Sanders, of Kentucky. It does not resemble the Maiden's Blush, as some have said, but is one of the very best Summer fruits. It is rather a good bearer, grows well, and a choice, fair fruit. It is a better apple than the Early Rose, but not quite so early. One of the very best. Two or three only earlier. "Fine southern apple."—*Trans. Ohio Pom. Society.*

BOUGH, *Early Sweet.* Color, yellow; form, roundish, conical ovate; size, 1; use, table; quality, 2; season, July and August.

REMARKS.—A fine sugary variety. The tree is a compact grower, and is rather a shy bearer. Subject to fall. When really ripe is fine, but never much flavor, beside the sweet. Exhibited by several members of the Cincinnati Horticultural Society. Committee on Fruit decide it "a fine saccharine variety."

BOXFORD.—Color, red striped; form, compressed; size, 2; use, table; quality, 2; season, August, or early in Sept.

REMARKS.—Superior to most. Tender, pleasant, but transient, and lacks character.

BRABANT BELLFLOWER. Color, yellow and red; form, roundish oblong; size, 1 to 2; use, kitchen; quality, 2 to 3; season, October to February.

REMARKS.—Sharp, sub-acid. Handsome and striking in appearance. Almost unworthy of cultivation, too much like Pennock. Its fine appearance will sell it in market, until, like many others of this kind, its real qualities, and those of better, are more known. "Little known in Ohio."—*Trans. Ohio Pom. Society.*

BRACKEN, synonymous with *White Juneating,* and *Juneating.* Color, pale yellow, sometimes a faint blush; form, flattish round; size, 3; use, table; quality, 2; season, early in July.

REMARKS.—Not very good for the kitchen. In southeastern Ohio a favorite early apple; also in Virginia; elsewhere, little known. An old foreign variety, popular in some sections for its earliness. "Resembles, if not identical with, Early Harvest, or Yellow Harvest" (Prince's). —*Trans. Ohio Pom. Society.* There is another apple near

Mr. Sayer's, Reading Road, thought distinct from Yellow Harvest.

Brenanan. Season, August and September.

REMARKS.—Regarded by the National Pomological Society as good.

BROADWELL SWEET. Color, light yellow; form, roundish, somewhat flattened; size, 2 to 1; use, table; quality, 1.

REMARKS.—Approved wherever it is known. A very fine, sweet Winter apple. A rather slender grower; young wood varicose. Specimens before the Cincinnati Horticultural Society, from Mr. Petticolas, last Winter, January, 1855. This variety is highly recommended for general cultivation. Its origin, Miami county, Ohio. Brought into notice here, first, by Mr. Wm. Resor. The best sweet apple. Lady's Sweeting not better, if so good.

Bronson's Sweeting.

REMARKS.—From Summit county, Ohio. "Not approved."—*Trans. Ohio Pom. Society.*

BULLOCK'S PIPPIN; see also American Golden Russet. Color, yellow, with soft russet; form, roundish ovate, tapering to the eye; size, 3; use, table; quality, 1; season, November to March.

REMARKS.—A very highly flavored apple; none more so. One of the finest dessert fruits in December, but must not be too ripe. The tree grows compactly, and is not an early bearer. This fruit should command double price, otherwise it is not profitable, and it does generally. A sure bearer and profitable; eaten at the proper point, a great fruit—the very best. Bears frost after they are picked better than most apples.

BUTTER APPLE, see *Apple Butter*.

REMARKS.—Many sweet seedling apples in the West have this name. This apple is little known, but good.

CABLE'S GILLIFLOWER (Baltimore, of Elliott, author of American Fruit Grower's Guide, and the nearest, except the present work, to the locality of Cincinnati, except Coxe). Color, light yellow, striped, and splashed with red, a little bronzed russet about the stem; form, round; size, 2 to 1; use, table and kitchen; quality, 2; season, November to February, and sometimes March.

REMARKS.—"Second-rate."—*Trans. Ohio Pom. Society.* This is not the Baltimore of Lindley, which is a pale lemon color, tinged with red, and with large open calyx.

Calville, White and Red.

REMARKS.—The Calvilles are not admired in this country; at least in the West.

CAMPFIELD. Color, green and yellow, with a reddish blush; form, roundish flattened; size, 2; use, cider; quality, 2; season, December.

REMARKS.—For stock and cider. One of the best and most productive sweet-cider apples.

CANNON PEARMAIN. Color, yellow with dull red, large yellow specks, russety; form, roundish, often angular; size 2; use, table; quality, 2; season, from December to March.

REMARKS.—"Good Winter."—*Trans. Ohio Pom. Society.* Valued much by F. G. Cary, College Hill, near Cincinnati. A good apple for baking in Winter, and very good to eat for those who like sweet apples.

CANADIAN PIPPIN, or *Canada Reinette*, with many other

names. Color, light greenish yellow; size, 1; use, table; quality, 1 to 2; season, November to April.

Remarks.—A great bearer, and of fine growth. Of more value in Canada, or at the North than around Cincinnati; still, it is a good fruit here. Some have thought it to be our favorite White Pippin, but it is not, as we have proved by having had specimens from the North at the Cincinnati Horticultural Society; and Dr. Warder saw unquestioned specimens in New York State, which were not our White Pippin, which was probably one of Wharton's introduction. It is certainly not our White Pippin, although very similar in many respects.

Carolina Winter Queen, or Nix's Apple. Color, yellowish green; form, flat, or nearly so; size, 2; use, table; quality, 2 to 3.

Remarks.—This apple is for the South, where it is very good. It will not suit us at Cincinnati.

Carolina Red June. Color, red; size, 2; use, table; quality, 2; season, July and August.

Remarks.—Large as Summer Queen. Cultivated much in the North; a very distinct variety; of medium size; deep, shining red and white; tender; flesh of pleasant flavor.

Carnel's Favorite.

Remarks.—Pronounced, "Very good," by the National Pomological Society.

CARTHOUSE, *Gilpin, or Romanite* (see both titles).

Remarks.—Is prolific, and good for cider. See the other names, for further description, of which it is well worthy.

Carter Apple.

REMARKS.—From Virginia. Not decided upon by the National Pomological Society.

CARY'S SEEDLING. Color, clouded green and yellow; form, round oblate; size, 2; use, table and kitchen; quality, 1; season, September and October.

REMARKS.—Originated at Farmers' College, seven miles from Cincinnati. Took premium at Ohio State Fair in 1852. Fine grained, tender, pleasant flavor; acid until quite ripe, when it becomes a pleasant sub-acid. Tree strong and vigorous, bears abundantly every second year, like some others, as the White Juneating, Reece's Apple of Campbell county, Ky., etc. This is a valuable variety.

Catshead, or Cathead Greening, or Round Catshead. Color, yellowish green; form, roundish; size, 1; use, kitchen; quality, 3; season, September and October.

REMARKS.—Not of much value, only for cooking. There is another, equally worthless, same shape, but striped yellow and red. "Second rate."—*Trans. Ohio Pom. Society.*

CATLINE, or *Gregson Apple.* Color, yellow; form, flat; size, 3; use, table; quality, 2; season, September and October.

REMARKS.—Productive, and bears early. Shoots, straight and delicate. "Good; little known. Of Eastern Ohio."—*Trans. Ohio Pom. Society.*

CHANDLER. Color, yellowish green; form, round, imperfect; size, 1; use, kitchen; quality, 2 to 3; season, October to January.

REMARKS.—Very acid. Exhibited by R. Dana to Cincinnati Horticultural Society, August, 1855; pronounced by Fruit Committee, "Second rate." "This must be an error of the Ohio Pomological Society; Chandler is a dark red apple, full of white spots."—*Dr. Mosher.*

Cheeseborough Russet, or Howard Russet, Kingsbury Russet, York Russet, of some, Forever Pippin, of some, West. Color, thin russet, or greenish yellow; form, conical; size, 1; use, kitchen; quality, 3; season, September and October.

REMARKS.—"Large, coarse, third rate."—*Trans. Ohio Pom. Society.*

CLARKE'S SEEDLING, or *London Sweet, London Winter Sweet, Winter Sweet, and Heicke's Winter Sweet.* Color, pale yellow; form, roundish, flattened; size, 2; use, table; quality, 2; season, November to February.

REMARKS.—Much cultivated here, South Ohio. Sometimes plentiful in market. This fruit somewhat resembles Broadwell. Well suited to the rich, deep soil of the Miami Valleys.

CODLIN, or *Royal Codlin.* Color, bright yellow; form, roundish, flattened; size, 2; use, kitchen; quality, 2; season, August.

REMARKS.—Exhibited often by R. Buchanan, Esq., at the Cincinnati Horticultural Rooms. It is ribbed a little. The flavor is sweetish, lively, and good. It is greenish near the stalk.

CODLIN, or *Keswick Codlin.* Color, greenish yellow; form, conical oblong; size, 2; use, kitchen; quality, for cooking only; season, July and August.

REMARKS.—A great bearer. Tree, round head, very hardy. The fruit cooks well in about ten or fifteen minutes; is tender, and acid. The tree bears when quite young, and the fruit hangs long and well on the tree.

"Popular Summer cooking apple."—*Trans. Ohio Pom. Society.* It is very similar to the Dutch Codlin. "The English Codlins coddle here as well as in their native land, where they have passed into a proverb for their supreme excellence as a stewing apple; but that called the English Codlin here, is often blotched and knotty."—*Trans. Am. Pom. Society.* It is not so, near Cincinnati. "Summer apples, of the best English sorts, invariably become too acid under our sun. (!) Red Astrachan, from North of Europe, is acid and dry."—*Trans. Am. Pom. Society.* Keswick Codlin, exhibited often by the author at Cincinnati Horticultural Rooms, and very fair and fine.

Clyde Beauty, or *Mackie's.* Color, pale greenish yellow, striped and mottled with light red, deep crimson in the sun; form, roundish conical, slightly ribbed; size, 1; use, table; quality, 2; season, September to December.

Remarks.—Deserves to be cultivated, from its fine-grained, juicy, sub-acid, very pleasant qualities.

Cole. Color, bright red; form, roundish, little conical; size, 1 to 2; use, table; quality, 2; season, July and August.

Remarks.—An early bearer, and a pleasant, sprightly fruit; also hardy. Fruit, juicy and "very good."

COOPER. Color, greenish yellow, with stripes and blotches of pale red; form, roundish, flattened; size, 1; use, table; quality, 1; season, August to November.

Remarks.—Best on rich, limestone clay. Rather subject to canker.° Flesh, yellowish, crisp, juicy. It is

° One sort of Canker proceeds, we think, from the ravages of the *white aphis*, or apple louse, which has the appearance of a white mold.

Another kind of Canker of apple trees may, in some situations, be caused by the uncongenial nature of the soil. A person has informed us excellent, and of the finest texture. Dr. Humphreys, of Portsmouth, Ohio, says it is from France. Barrels of Cooper apples were exhibited at the Ohio State Fair, in 1850, and nothing in season equaled them. It is, certainly, one of the very best Fall apples. Professor F. G. Cary says, "It is exceedingly well adapted to the locality of Cincinnati." The tree is a fine, strong, upright grower. The fruit is fair and beautiful.

COOPER'S REDLING, of New York.

REMARKS.—"Little known; second rate."—*Trans. Ohio Pom. Society.*

COOPER'S RUSSETING. Color, yellow, with some russet; form, long ovate; quality, 2; season, November to Spring.

REMARKS.—Fruit, dry, sweet, and rich. Excellent for cider, and cooking.

Cornish Gilliflower. Color, dark green and yellow; form, ovate; size, 2; use, table; quality, 3; season, August to April.

REMARKS.—Much esteemed in England. A bad bearer

---

that having observed many of his apple trees become cankered at a certain period of growth, he was induced to examine the nature of the soil, at the greatest depth the roots had penetrated, and which he found consisted of gravel. Not being willing to give over the propagation of apple trees, he caused a pavement of bricks to be made on the bed of gravel, which obliged the roots to take a horizontal direction, and thereby prevented them from reaching the gravel, since which they have been free from canker.

Another kind of Canker, or Blight, is a sort of white mold around the lower part of the trunk, which causes the bark to part from the wood; probably caused by the Winter sun (often particularly warm, even at that season of the year, in this climate) after severe frosts, which, when the sap has been frozen causes it to thaw, and being unable to circulate, it goes into fermentation, and rots.

there, but promises better here. It is very like the Red Gilliflower. Both are poor.

CRAB APPLES. See the varieties; chiefly for cider, or preserves; the largest for cider, the smallest (as the beautiful red Siberian) for preserves.

CRACKING. Color, yellowish white; form, roundish; size, 1; use, table; quality, 1; season, September to January.

REMARKS.—Of Eastern Ohio. Tree, strong grower; requires little pruning. Flesh, tender, juicy. "Very good." New. "Large and showy; early Winter."—*Trans. Ohio Pom. Society.* Very fine wherever known. Highly esteemed about Massillon, Ohio.

Culp. "A seedling of Jefferson county, Ohio."—*Trans. Ohio Pom. Society.*

CUMBERLAND SWEETING, or *Cumberland.* Form, roundish, flattened; size, 1; quality, 2.

REMARKS.—Valuable for market.

DANVERS' WINTER SWEET. Color, dull yellow, and orange; form, roundish, oblong; size, 1 to 2; use, kitchen; quality, 2; season, all Winter.

REMARKS.—Cooks very well. Flesh, yellow, firm, sweet, and rich. Excellent for stock. Exhibited by the author, August, 1855, at the Cincinnati Horticultural Society's Rooms. Not equal, in the West, to Sweet Butter, Lady's Sweeting, nor Broadwell.

DANA'S BAKER.

REMARKS.—"Of Washington county, Ohio. Generally approved."—*Trans. Ohio Pom. Society.*

DANIEL.

REMARKS.—Of Zanesville, Ohio. "Little known, but good."—*Trans. Ohio Pom. Society.*

December Russet.

REMARKS.—Of Morrow county, Ohio. "Little known." —*Trans. Ohio Pom. Society.* A fine keeper.

DELIGHT. Color, yellow and red; size, 2; use, table and kitchen; quality, 2; season, October to March.

REMARKS.—"A good and beautiful apple."—*Elliott.* From R. Buchanan's orchard—an orchard which has largely contributed to the beauty and interest of the Cincinnati and other Societies' Exhibitions.

DETROIT. See White Bellflower.

REMARKS.—One of the best apples grown.

DOCTOR. Color, yellow with red; form, regular, flat; size, 1; use, table; quality, 2; season, September to February.

REMARKS.—Much grown in Southern Ohio and Indiana. Not *generally* much grown, for though a pleasant and handsome, crisp, juicy fruit, of large size, it is a poor bearer. The apple grown generally for this, hereabout (Cincinnati), is Hay's Winter Wine.

DOMINE. Color, red and green; form, regular; size, 2; use, table and kitchen; quality, 2; season, all Winter, or October to April.

REMARKS.—Origin not exactly known. Tree, a strong, vigorous grower. A tremendous bearer. A good household Winter fruit. A large, showy, fair fruit, when not too full.

DRAP D' OR, or *Bay Apple.* Color, dull yellow and gold; form, roundish flattened; size, 1; use, table and kitchen; quality, 1; season, August to November.

REMARKS.—One of the finest Autumn fruits. Ripens and falls during two months, and even when green is fine for cooking. "Most fair and delicious."—*Fruit Committee Cincinnati Horticultural Society.* "Excellently well adapted to the locality of Cincinnati."—*F. G. Cary*, President of the Society for the year 1856.

DUMPLING, or *Crooked Limb Pippin, French Pippin of Indiana, and Watson's Dumpling.* Color, light yellow, blush in sun; size, 1; use, kitchen; quality, 2; form, roundish, oblong; season, October to December.

REMARKS.—"Second rate."—*Trans. Ohio Pom. Society.*

DUTCH MIGNONNE. Color, dull orange; form, roundish, regular; size, 1; use, kitchen; quality, 2; season, October to February.

REMARKS.—This is one of the few great acquisitions from abroad. Its appearance is superb and flavor rich. The tree bears fine crops.

DYER, POMME ROYAL, OR SMITHFIELD SPICE. Color, pale yellow or white; form, globular; size, 1 to 2; use, table; quality, 2.

REMARKS.—Admired where known. An excellent variety, ripening in September, in this locality. An early bearer.

DUCHESS OF OLDENBURGH. Color, striped red and yellow; form, roundish; size, 2; use, table and kitchen, more of the latter; quality, 2; season, August and September, sometimes July, in Cincinnati.

Remarks.—Most beautifully striped, a kitchen apple almost entirely. Tree vigorous, with upright shoots.

EARLY HARVEST, *Princes', or Yellow Harvest.* Color, pale light yellow; form, round; size, 2; use, table and kitchen; quality, 1; season, July and August.

Remarks.—Most excellent for cooking. Well known and everywhere approved. "Excellently well adapted to the vicinity of Cincinnati." F. G. Cary, College Hill. Exhibited continually during season, and considered one of the very best early apples for all purposes except keeping. Some do not consider it equal to Early Yellow June (a counterfeit of Early June), which is less tart, and often planted for Early Harvest, and believed to be the true variety by many of the best pomologists. M. McWilliams, of the Cincinnati Horticultural Society Fruit Committee, has both. Manure with plenty of lime and potash. The Early Harvest of the late Dr. Flagg differs from Prince's Early Harvest.

Early Red Margaret. Color, green, red in sun; form, roundish, oblong, conical; size, 2 to 3; use, table; quality, 2; season, July.

Remarks.—Tree moderate bearer, with upright downy shoots. It is not equal to Early Strawberry.

Early Chandler, see *Chandler.* A favorite at Marietta, Ohio, and of G. Dana. "Handsome, high flavored, acid."—*Trans. Ohio Pom. Society.*

EARLY STRAWBERRY, or *Red Juneating.* Color, yellowish white, striped and stained over with bright and dark red; form, roundish, varying to conical and angular; size, 3; use, table; quality, 1; season, July.

Remarks.—Tree very erect, with dark wood. Good in

all soils and generally esteemed. Exhibited by many members of Cincinnati Horticultural Society. "A great favorite," Fruit Committee. "Excellently well adapted to the locality of Cincinnati." F. G. Cary, President for 1856.

Early Red Sweeting. Color, light red; form, roundish, flattened; use, table; quality, 3; season, August.

REMARKS.—Not approved generally.

EARLY JOE. Color, pale yellow and green; form, roundish, flattened; size, 3; use, table; quality, 2; season, August.

REMARKS.—Has a most delicate pear flavor. Not suited to orcharding, and of slow growth. It is very hardy, but should have a rich, strong soil. "Very delicate and excellent when in perfection."—*Trans. Ohio Pom. Society.*

EARLY JUNE, *July Pippin, Large White Juneating, Early French Reinette, etc.* Color, pale light yellow, with a few dots of white; form, roundish, rarely a little flattened; size, 2; use, table and kitchen; quality, 1; season, July.

REMARKS.—"Resembles very much Early Harvest and Bracken."—*Trans. Ohio Pom. Society.*

EARLY SWEET BOUGH, see *Early Bough.*

English Russet. Color, light green and yellow; form, roundish ovate, or conical; size, 2; use, kitchen only; quality, 3; season, all Winter and Spring.

REMARKS.—There is a Golden Russet like this in Central Ohio. A very long keeper, over a year; therefore, on this account, perhaps of some value, but it is not superior, not even for the kitchen. Decidedly poor, though prolific, and a great keeper. It is, at all events, not better than the American Pippin or Grindstone, which will also keep till July, and be no better after all than "dried apples," and often not near so good as many of them.

Early Pennock, Sleeper's Yellow, Warren Pennock, etc. Color, greenish yellow; form, roundish, tapering to the eye; size, 1; use, table; quality, 2 to 3; season, July and August.

REMARKS.—Is from Harrison and Belmont counties. Trees, vigorous, hardy, and prolific bearers. A little less than second quality.

EARLY RED STREAK. Color, striped red; form, roundish; size, 2; use, kitchen; quality, 3; season, August and September.

REMARKS.—An English cider apple merely. It thrives well in this country. It is one of the most coarse and tart of early apples. Much grown about Philadelphia and Cincinnati, and sells remarkably well.

ERNST'S SWEETING. Remarks.—"But little known."—*Trans. Ohio Pom. Society.* See Sweeting, Ernst's.

ÆSOPUS SPITZENBURG. Color, rich, lively red on yellow; form, flat at base, oblong, tapering to the eye, ribbed; size, 1 to 2; use, table; quality, 1 to 2 (near Cincinnati); season, January to March.

REMARKS.—Tree has a drooping head; stem varies in size. Handsome and very good, and highly approved in some localities, particularly toward the North. It requires much lime and potash. In the North is its place for superiority; and it is, there, exquisitely flavored. One of the best cooking apples in the catalogue. In Northern

Illinois it is subject to blight. In Southern Ohio and Indiana it is too large and less solid, and is often russety. It, there, ripens too early for a Winter fruit, but still fine.

Eve Apple, of the Irish, or Early June, Eggtop, etc., etc. Color, green, red in sun; form, roundish, oblong, conical; size, 3; quality, 2 to 3; season, July.

REMARKS.—An inferior fruit. Its shape is very long and singular.

FAHNESTOCK'S SWEETING.

REMARKS.—"Handsome shaped Summer apple."—*Trans. Ohio Pom. Society.*

Fall Bough.

"Little known, not approved."—*Trans. Ohio Pomological Society.*

FALL PIPPIN, *sometimes called Golden Pippin.* Color, green and yellow; form, roundish, conical, somewhat flattened: size, 1; use, table; quality, 1; season, September to December.

REMARKS.—Extensively grown in the West. Tender, sub-acid, aromatic. "Not Holland Pippin. Large, handsome and good."—*Trans. Ohio Pom. Society.* Is good about Cincinnati, and highly esteemed. It ripens gradually. Does not keep long. The Holland Pippin often confounded with it; but as the Ohio Pom. Society states above, not the same. Downing wonderfully suggested that these apples might be confounded. They are not at all alike. The Fall Pippin drops badly from old trees. More open at the eye than the Holland Pippin. Very good. "Excellently well adapted to the locality of Cincinnati."—*F. G. Cary*, President Cincinnati Horticultural Society for 1856.

Fall Heavy.

Remarks.—Little known here (Ohio). "Not as good as Fall Pippin."—*Trans. Ohio Pom. Society.*

Fallawater, *or Fallenwalder, or apple of the fallen timber, called, also, Talpahocken, from the creek of that name.* Color, pale yellow and green; form, roundish flattened, and roundish ovate; size, 1 to 2; use, table; quality, 2; season, December to May.

Remarks —The English books say this apple possesses the Newtown Pippin flavor. Ours does not at all. "Second rate."—*Trans. Ohio Pom. Society.* Valuable for distant markets. Has a thick skin. Requires rich limestone soils. Fruit, always fair and large. "Is a great bearer of very fair, large apples, wanting much flavor."—*Reports from Pennsylvania to the American Pom. Society.* In our own locality (Cincinnati), it is not so valuable for carrying far, as our soil seems to give it a flesh more soft, and, therefore, more easily bruised. Its flavor, even here, will suit people whose tastes are rather easily pleased. Its flavor is anywhere not high, and when cooked it has hardly any. Still it is a tolerably passable Winter fruit.

FALL QUEEN, *Red Gloria Mundi, Horse Apple, or Mundy's.* Color, green and yellow to orange, mostly striped with red; form, roundish conical; size, 1; use, table and kitchen; quality, 1; season, July and August to October.

Remarks.—Well known in Kentucky. Valued highly South and West, especially for cooking. "Of Southern Ohio. Large and handsome."—*Trans. Ohio Pom. Society.* A grand apple for early Winter. One of the best large apples, worth fifty Gloria Mundi's, and very superior to the above apple—Fallawater.

Fall Wine. Color, rich red, green at stem; form, rounded, flattened; size, 2; use, fine table; quality, 2; season, September to October.

Remarks.—This is distinct from the Wine Apple.

Father Abraham, from St. Louis.

Remarks.—"Wood-thrifty, but thorny. Fruit, good; form, conical; color, yellow, with a red side; flavor, high; a good keeper." *Fruits of Missouri*, by Thomas Allen, of St. Louis. "A fine table apple. Little known."—*Trans. Ohio Pom. Society*. Exhibited at Cincinnati, at the Horticultural Society Fair in 1843. It was then and there regarded valuable. It is not a great bearer, but regular. Of *fenoullet* flavor, which becomes rather tiresome, though agreeable at first.

Fameuse, or *Pomme de Neige*, or *Snow Apple*. Color, crimson red, with a greenish yellow; form, round; size, 2; use, table and kitchen; quality, 2; season, September and October to December.

Remarks.—Of uncommon beauty. Flesh, snowy white, tender and delicious. Succeeds best in the North. A most delicate Canadian or Northern Winter fruit, but with us a Fall apple, very prolific, beautiful, delicate flesh, juicy. Of no decided flavor; still less when cooked, but cooks well.

Father Apple. Size, 2; use, kitchen.

Remarks.—From West Bloomfield, Ontario county, New York. Not yet tested in this locality. J. R. Watts, Esq., of Rochester, says the Father Apple is worth its weight in gold. It makes a fine sauce without sugar, and cooks well.

Father Abraham. Color, red, little yellow, spots and

blotches of darker red; form, flat; size, 2; use, table and kitchen; quality, 2; season, early Winter.

REMARKS. — From Virginia. Considerably grown in Kentucky. Coxe esteemed this apple of much value in his time. It keeps till April.

FAVORITE.

REMARKS.—"Of Kentucky. Little known in Ohio.—*Trans. Ohio Pom. Society.*

Federal.

REMARKS.—"Of Mr. Springer, of Ohio. Little known." —*Trans. Ohio Pom. Society.*

FINK.

REMARKS.—"A seedling of Perry county, Ohio. Very long keeper. Good."—*Trans. Ohio Pom. Society.*

Fleiner. Color, striped, lemon yellow and red, red cheek, oblong; size, 2; use, table; quality, 2; season, September and October.

REMARKS.—Its great productiveness is its chief value, which is very remarkable.

Flower of Kent. Color, dull yellow, or tawny, and red; form, roundish flattened, use, cooking and baking; size, 1; quality, 2 to 3; season, September to January.

REMARKS.—Coarse, sub-acid, and hardly worthy of cultivation.

FLUSHING SPITZENBURG. See *Newtown Spitzenburg, Ox Eye,* and *Joe Berry.*

FORT MIAMI. Color, brownish red, more or less russeted; form, roundish oblong, flattened at both ends, widest at base, uneven, somewhat ribbed; size, 1 to 2; use, table; quality, 2; season, January to April.

REMARKS.—Flavor highly spicy, with a sub-acid taste. Does not bear early, but when older produces fine crops.

FRANKLIN'S GOLDEN PIPPIN. Color, pale light greenish yellow, with white finish net-work, few russet dots; form, roundish ovate, regular conical; size, 2; use, table and kitchen; quality, 1; season, September.

REMARKS.—A good, rich, yellow green apple, but little known. From Dr. Hildreth of Marietta. One of the nicest apples of the season. Comes in just before the Bellflower. Rich, juicy, good, regular bearer. Grown near Mr. Sayers, Reading Road.

FRONCLIN. Color, bright red; form, regular round; size, 2; use, table and kitchen; quality, 2.

REMARKS.—"Very prolific, even when young."—*Pennsylvania Pom. Trans.*

FRENCH PIPPIN, see *Newark Pippin*, the same apple, one of the most valuable of fruits, only rather feeble in growing. There is another, of little value, of this name.

GABRIEL, or *Ladies' Blush*. Color, yellowish striped, and splashed with pale red; form, roundish conical; size, medium or 2; use, table; quality, 2 to 1; season, October and November.

REMARKS.—"An excellent Fall apple."—*Trans. Ohio Pom. Society.*

GATE or *Waxen*, see *Belmont.*

GENNETING.

REMARKS.—Of Muskingum Valley, Ohio. "Early, acid,

second-rate."—*Trans. Ohio Pom. Society.* This is not the valuable Raules' Janet, or Genneting, of Kentucky.

GEORGE APPLE. Color, pale yellow; form, roundish flattened; size, 2; use, table; quality, 2; season, July.

REMARKS.—" Resembles Early Harvest, perhaps identical."—*Trans. Ohio Pom. Society.*

GILPIN, *Romanite of the West,* or *Carthouse.* Color, red and yellow; form, round flattened; size, nearly 3; use, table and cider; quality, 2.

REMARKS.—Tree very hardy and prolific. Generally escapes frost. Worth double the price of other apples in general, in June. Very valuable, therefore, as an orchard market fruit, but of nearly third-rate quality. Mr. Sam'l Carter, two miles back of Newport, Kentucky, considers it his most profitable market fruit in the month of June. It makes good cider, and is rich in giving body to others. Bears bruising remarkably well, may be knocked off with poles, shoveled into a cart, dumped down on the ground, and holed away like turnips, and go to market, next Spring, in sound condition. " Rather small, good keeper, second rate as to quality."—*Trans. Ohio Pom. Society.*

Gloria Mundi, Monstrous Pippin, American Mammoth, Baltimore Pippin, Pound, etc. Color, lemon yellow, dull white spots; form, round, rather angular, flattened; size, 1; use, kitchen; quality, 3; season, September to January.

REMARKS.—Gigantic. Sometimes weighs near a pound and a half. It is pretty good for cooking and drying. The tree is vigorous, and bears well. "Very large, second rate."—*Trans. Ohio Pom. Society.* Exhibited often before the Cincinnati Horticultural Society, and monstrous specimens. Generally considered "unworthy," except for curiosity of size and for show. The Talpahocken, or

Fallawater, often approaches it in size, but is a much better apple. Falls a great deal in rich soils. The tree becomes very large. It succeeds well with us, except on clayey soils, and deep prairies.

Gloucester White

REMARKS.—"Not deserving to be known."—*Trans. Ohio Pom. Society.*

Glory of York. See Ribston Pippin, the finest apple of England, and of the Northern States.

GOFF APPLE.

REMARKS.—"A seedling, from Champaigne county, Ohio. A good apple."—*Trans. Ohio Pom. Society.*

GOLDEN.

REMARKS.—Of Perry county, Ohio. "Resembles Early Harvest; perhaps identical."—*Trans. Ohio Pom. Society.*

GOLDEN DROP. See Court of Wyck.

GOLDEN BALL, *Belle et Bonne (called Belly Bound, a corruption of Belle et Bonne), or Connecticut Apple.* Color, rich yellow, sometimes a faint blush near the stalk, and with rough dots; form, roundish, lessening to the eye; size, 1; use, table and kitchen; quality, 2; season, September to December.

REMARKS.—Pretty good. This is not the same as the Belle et Bonne of Lindley. It requires a rich, strong, and compact, or heavy soil. It bears well as it grows older.

Golden Harvey.

REMARKS.—"Unworthy."—*Dr. Warder*, one of our best Western Pomologists.

Golden Pippin.

Remarks.—Rather a vague name in this country. Once very good in Britain, but of late years failing there; in some places becoming very hard and specky, and fine, again, in other parts. In America it is a synonyme, erroneously, for Fall Pippin, Ortley, and several other yellow apples. Some of the fruit, under this name, is unworthy.

Golden Reinette.

Remarks.—"Inferior; foreign."—*Dr. Warder.* "A fine *European* Dessert Apple."—*Trans. Ohio Pom. Society.*

GOLDEN SWEET. Color, pale yellow; form, round; size, 1; use, kitchen; quality, 1; season, August and September.

Remarks.—One of the best baking fruits, without sugar. On list for stock. "Good and profitable Summer."—*Trans. Ohio Pom. Society.*

GOLDEN RUSSET. See American Golden Russet, Bullock's Pippin, etc.

Governor. Of Columbus, Ohio.

Remarks.—"Little known."—*Trans. Ohio Pom. Society.*

Gravenstein. Color, bright yellow; form, rather flattened, one-sided; size, 1; use, table, but chiefly cooking; quality, 2 to 1; season, August and September.

Remarks.—A German fruit. Bears well young. A strong grower. "Handsome, good, early Summer."—*Trans. Ohio Pom. Society.* Sometimes excellent. Probably not generally profitable. Sometimes very coarse, and too acid.

GREEN NEWTOWN PIPPIN. See Newtown Pippin, Green.

GREEN'S CHOICE.

REMARKS.—Handsome; red striped; passed by the National Pomological Society, "Very good."

Green Sweeting.

REMARKS.—"Of New York. Little known in Ohio." —*Trans. Ohio Pom. Society.*

Grey House. See Black Vandevere.

GRIMES'S GOLDEN PIPPIN.

REMARKS.—"A good seedling from Virginia."—*Trans. Ohio Pom. Society.*

Grindstone. See, also, American Pippin. Color, red on green; form, flat; use, kitchen; quality, 4; season, all Winter, and till July.

REMARKS.—Abundant bearer. Hangs well. Like the Lansingburgh, Virginia Greening, and a few others, a very great keeper, but a very poor apple, almost worthless.

Gully Apple.

REMARKS.—Pennsylvania apple; much boasted of. Little known here, at present. It would be well to try it, and all others with a great reputation elsewhere.

HAGLOE. Color, red; form, round; size, 2; use, table and kitchen; quality, 2; season, July.

REMARKS.—Distinct from Hagloe Crab, a small, ovate, cider apple. (See title, Cider Apples, in our descriptions of apples.) Exhibited by A. H. Ernst.

HARRISON. Color, green and yellow; form, conical; size, 3; use, cider; season, Winter.

REMARKS.—Very good cider apple (see Cider Apples).

The Jersey cider apple for Champagne. Prolific. Holds well. May be gathered late, and manufactured any time. It keeps long, and is a good cooking apple near Spring.

Hartford Sweeting. Color, red striped on greenish yellow; form, roundish, flattened; size, 1; use, table; quality, 2; season, November to March.

Remarks.—Flesh, whitish, tender and juicy. "Very good."

Hawley.. Color, pale yellow and green; form, round, varies; size, 2; use, table; quality, 2; season, Fall.

Remarks.—Fine flavored. Flesh, yellowish white, tender, with a rich sub-acid. "Excellent."

Hawthornden. Color, pale light yellow, blush in the sun; form, pretty regular, roundish, rather flattened; size, 2 to 1; use, table, cooking or drying; quality, 2; season, Autumn.

Remarks.—A Scotch apple. It has some resemblance, but is not equal to that good, and to most, agreeable apple, the Maiden's Blush. It cooks as well. Is little known in Ohio; grown more in the North-west, where it is more suited. It has not yet been tried in this locality—at least we have not seen it. If it is as good as Maiden's Blush, it will do.

Hay's Winter, or *Wine Apple.* Color, pale green, becoming yellow when near ripe; form, roundish flattened; size, 2 to 1; use, table and kitchen; quality, 2; season, December to March.

Remarks.—Much grown in Ohio and Indiana. Unworthy its high repute among some. Confounded with Dr. Dewitt.

HEICKE'S SUMMER QUEEN, of Dayton, Ohio. "A good apple."—*Trans. Ohio Pom. Society.*

HEICKE'S WINTER SWEET, *London Sweet,* and *Clark's Seedling;* see *London Sweet.* Plentiful in Cincinnati market. Resembles that most valuable fruit, the Broadwell. Adapted to the rich, deep soils of the West, like the fair, rich and beautiful Broadwell.

HELEN'S FAVORITE. Color, dark red; form, round; size, 2; use, table; quality, 2; season, January and Feb.

REMARKS.—From Troy, Ohio. "Very Good."—*A. H. Ernst's Reports.* "Very good."—*Dr. Warder.*

HEREFORDSHIRE PEARMAIN, *Royal Pearmain, Old Pearmain,* etc. Color, brownish red, mottled with russety green; form, roundish, conical; size, 2; use, table; quality, 2; season, December to February.

REMARKS.—Requires a rich, strong soil; fruit then is fine, and of superior excellence. Seldom seen here. Grown in New York and Northern Indiana, and Illinois. It is there very rich and fine. We do not think the apple we have by this name is correct, as it does not answer the description in the books.

HEWES' VIRGINIA CRAB. Color, dull red, dotted with white specks, and obscurely streaked with greenish yellow; form, nearly round; size, 4 to 5; use, cider; quality, 1; season, Winter.

REMARKS.—Very prolific, and very fine for cider—surpassingly fine. Cider from it keeps very long. Tree very hardy, though small.

HETERIC. Form, regular round; size, 2 to 1; use,

table and kitchen; quality, 2; season, November to January.

REMARKS.—A Pennsylvania apple. A decided favorite. Fine bearer. Form always perfect.

HERMAN. Color, red; form, oblong; size, 2; quality, 2.

REMARKS.—Flesh rather greenish, tender, juicy, and of high flavor.

HOLLAND PIPPIN. Color, greenish yellow; form, more regularly round than the Fall Pippin, stem shorter, and cavity not so deep, more broad and open; size, 1; use, kitchen; quality, 2; season, fit for pies in August to January.

REMARKS.—Often confounded with the Fall Pippin. Different, however; does not approach to its excellence. The Gravenstein is better even for cooking. The time of ripening is also different. It deserves a place in the garden. Sometimes called Golden Pippin. Superior for cooking and drying, and a good table apple. "Not Fall Pippin."—*Trans. Ohio Pom. Society.*

HONEY GREENING, or *Green Sweet.* "Of Washington county, Ohio."—*Trans. Ohio Pom. Society.* Color, dull greenish white, or yellow, with greenish white, or sometimes pale russet dots; form, roundish; size, 2 to 1; use, table; quality, 2; season, September and October.

REMARKS.—Is a large fruit in Indiana, as, indeed, most of our apples are. Flesh, greenish white, tender, juicy, and quite sweet.

HOOPER'S RED STREAK. Form, round; size, 3; use, table and cider; quality, 2; season, October.

REMARKS.—A very handsome Fall apple. A Seedling. Small, remarkably juicy, and very sweet for a sub-acid apple: unequaled, indeed, in this respect. Tree very large and straight, with long, upright branches.

HORSE APPLE, a synonyme of *Fall Queen*, which see.

Horse Apple, of T. V. Petticolas. "Not approved."—*Trans. Ohio Pom. Society.*

HOWARD.

REMARKS.—Regarded as very good by the National Pom. Society. Beautiful and excellent, North and East.

Hubbardston Non-such, or *Sutton*, etc. Color, yellow ground, striped with rich red; form, roundish ovate; size, 1; use, table and kitchen; quality, 2; season, October to June.

REMARKS.—Very like Baldwin and Red Canada, but larger. Worthy of orchard culture in suitable places, but not in Southern Ohio. It ripens, here, too soon, and drops badly. Is too coarse for table or dessert. Baldwin has been widely distributed under this name. The tree is spreading and productive.

Jack Apple, or *Early Jack*. Color, deep brown, with green spots.

REMARKS.—When ripe the fruit is almost white, very juicy, and of such an undecided sweet and sour, that it can not be said to be either.

Jersey Black.

REMARKS.—"Little known, and of little value."—*Trans. Ohio Pom. Society.*

Jersey Red Streak, of Mr. Springer. "Not approved."—*Trans. Ohio Pom. Society.*

Irish Peach Apple. "Unworthy."

IMPERIAL VANDEVERE.

REMARKS.—"A Pennsylvania apple. Seedling of the Smoke House, but rather coarser." *Pennsylvania Pom. Reports.*

Ironstone, or *Eastern Pippin.* Color, deep green, brownish blush; form, roundish; size, 2; use, kitchen; quality, 3; season, November to June.

REMARKS.—Sprightly, juicy, and pleasant.

JEFFRIES.

REMARKS.—Regarded by the National Pomological Society as "Best." Season, August and September; color, pale yellow, striped and stained with red, becoming dark in sun; form, roundish flattened; size; 2; use, table; quality, 1.

JENKINS. Color, red, with large white dots on a yellowish ground; form, roundish ovate; size, 3; use, table; quality, 2; season, Winter.

REMARKS.—Flesh, white, tender, as good, perhaps better than the famous little Lady Apple. Beautiful and excellent for evening entertainments. But whether, like the Lady, it will deteriorate somewhat, West, remains to be seen.

JOE BERRY, a synonyme of *Newtown Spitzenburg,* or *Ox Eye.* See Newtown Spitzenburg, Vandevere, of New York. Color, red on yellow; form, round, flattened; size, 1; use, table; quality, 1; season, December to January.

REMARKS.—This is strangely confounded with Vandevere by Downing, but it has been unquestionably identified as the Newtown Spitzenburg of Coxe, and that confirmed by one of his own children, Mrs. McIlvaine of Clifton, near Cincinnati. A superior, rich, spicy apple, without the acidity of the Æsopus. It is, however, prone to fall from old trees.

Johnson's Sweet.

REMARKS.—" A seedling, from Harrison county, Ohio." —*Trans. Ohio Pom. Society.*

JONATHAN, *King Philip, Wine, Winesap* (erroneously). Color, light yellow ground, mostly overspread, streaked or stained with rich, light red, and with a few minute light dots; form, roundish conical, regular; size, 2; use, table; quality, 1; season, November to January.

REMARKS.—Requires a rich, strong soil. A New York apple. Flesh, tender, juicy, and rich, with a good deal of the Spitzenburg character; shoots, light colored, slender, and spreading. Rather a feeble grower, and indifferent bearer, even in its best localities, but particularly with us, West. It is a very fine and attractive apple when shown on the stalls, from its rich red color.

JERSEY SWEETING. Color, red; form, oblong; size, 2; use, table and kitchen; quality, 1; season, August and September.

REMARKS.—" First rate for table; early Fall."—*Trans. Ohio Pom. Society.* A beautiful, tender, sweet apple; ornamental as well as good. Of good baking qualities. To follow Early Bough, in the hog, or stock list.

JUNEATING, WHITE. Color, pale green, then light yellow; form, round, a little flattened; size, 2 to 3; use, table and kitchen; quality, 2; season, July.

REMARKS.—Very old. Mentioned by Evelyn, in 1660; by Ray, in 1688. A very common apple, in most gardens.

Has a fine scent. Apt to crack a little, and become dry. Should be used immediately, when ripe. An abundant bearer on alternate years. The earliest on the hog, or stock list.

Kaighn's Spitzenburg, Long Pearmain, Lady Finger, Long John, Red Pearmain, etc. Color, yellow, mostly overspread with bright, clear red; form, oblong, rounded; size, 1; use, table and kitchen; quality, 2 to 3; season, October, November, and December.

REMARKS.—This fruit is large and showy, and that is about all that can be said of it.

KEISER.

REMARKS.—"Of Coshocton, Ohio. Good, not first rate." —*Trans. Ohio Pom. Society.*

Kentish Fillbasket. Color, yellowish, brownish blush; form, roundish and ribbed; size, 1; use, kitchen only; quality, 3; season, September to January.

REMARKS.—Flesh tender, but not a desirable fruit.

Kenrick's Autumn. Color, pale yellow; form, round, flattened at base; size, 2; use, kitchen; quality, 2 or 3; season, August and September.

REMARKS.—Flesh white, a little stained with red. Not worthy.

KESWICK CODLIN. (See Codlin, Keswick.)

REMARKS.—A fine, fair, pure-complexioned cooking apple, requiring sugar; of good size, and saleable in market on account of its earliness (July), and tempting appearance. Hangs for two months, or more, on the tree. Not fit for table, only for cooking.

Kerry Pippin.

Remarks.—"Deemed unworthy of cultivation."—*Trans. Ohio Pom. Society.*

KING PHILIP. (See Jonathan.)

Kirke's Lord Nelson. Color, red on pale yellow; form, roundish; size, 1; use, kitchen; quality, 3; season, October to December.

Remarks.—Foreign. Of but little value here.

Kilham Hill. Color, pale yellow and red; form, roundish, ribbed; size, 1; use, kitchen; quality, 3; season, September.

Remarks.—Origin, American; very questionable value.

King of the Pippins. Color, yellow and red, splashed; form, roundish oblong; size, 2; use, kitchen; quality, 3; season, September and October.

Remarks.—In the West, sometimes confounded with Gravenstein.

Knight's Golden Pippin, Downton, or Elton Pippin. Color, yellow; form, roundish; size, 3; use, table and kitchen; quality, 3; season, September and October.

Remarks.—American origin—but inferior.

Kingsley. Color, yellowish, striped and splashed with pinkish red, some white dots; form, roundish, oval; size, 2; use, table; quality, 2; season, October to May.

Remarks.—"Best."—*J. C. Wendell, in Pom. Trans.*

King Apple, from Western New York. Color, pale yellow ground, with two shades of red, striped and splashed, brown dots and russet; form, roundish, oblong, somewhat ribbed; size, 1; use, table and cooking; quality, 2; season, October to December.

KING APPLE, from Mason county, Kentucky. Color, yellow, clouded with dull red, some small specks, moldy or fungus-like patches often found in Western and Southern apples; size, 2 to 1; use, table and baking; quality, 2; season, November to January, and even February.

KANE. Color, white, with red cheek; form, flattened, round; size, 2; use, table; quality, 2; season, Autumn.
REMARKS.—Resembles Maiden's Blush, but not so flat. Very beautiful. From Philadelphia.

KEIM. Color, fine pale yellow, waxen; form, rounded, inclining to conical; size, 3; use, table; quality, 2; season, January to April.
REMARKS.—"Very good. Flavor mild and pleasant."—*Horticulturist, Philadelphia.*

Knight's Codlin. Color, greenish yellow; form, roundish; size, 2; use, cooking; quality, 3; season, September to November.

KOHL APPLE.
REMARKS.—A variety imported from Germany. The tree is thrifty; regular top; good and regular bearer; fruit small, regularly shaped, crisp, juicy, and well flavored—a good keeper. *Fruits of Missouri,* by Thos. Allen, of St. Louis.

KREMLIN.
REMARKS.—Exhibited by Mr. Fee, August, 1855. Fruit Committee decided it to be "a very large, early apple,

tender flesh, and very valuable for its early ripening, and other good qualities."

Ladies' Blush. Color, yellow, with red blush; size, 2.

Remarks.—American origin. Flesh too dry. The *Gate* was exhibited for some time, under this name, at Cincinnati Horticultural Rooms, by J. C. Ferris. It is a synonyme of *Gabriel.*

Lady Finger, see Kaighn's Spitzenburg, of which it is a synonyme. Also, the name of a very poor apple in Pennsylvania; seen, also, in Central Ohio. Unworthy.

LADIES' SWEETING. Color, red in the sun, pale yellowish green in the shade, with broken stripes of pale red; form, roundish ovate; size, 1; use, table; quality, 1; season, keeps till May.

Remarks.—A favorite in New York. Not considered equal, there, however, to Danvers' Winter Sweet. "Handsome and good, second rate."—*Trans. Ohio Pom. Society.* Here, it is one of our best sweet apples. Yellow flesh. Winter. Perfumed like an orange, though not so strong.

LADY APPLE, or *Api, Pomme Rose, Petit Api, etc.* Color, bright red, on clear yellow; form, flat; size, 4; use, table and cooking (being as rich as a preserve); quality, 1; season, November to May.

Remarks.—Tree very upright, like the cherry. Fruit in beautiful clusters. It takes ten years before very productive. When it bears small, perfect fruit, one of the most profitable in the East, selling at ten dollars per barrel; but in the West, it is not so much appreciated as an ornament of the table, nor does it grow so fine and sightly. Small and beautiful. "Profitable in some localities."—*Trans. Ohio Pom. Society.*

Lancashire Witch.

REMARKS.—Handsome. A profitable apple, not much grown here, yet ought to be.

Lancaster Greening. Color, greenish yellow, with brown dots and blotches; form, regular roundish conical; size, 2; use, table and kitchen; quality, 2; season, all Winter to April.

REMARKS.—Valued as a long keeper. Remains good on tree till very late. Flesh, greenish, juicy, and sub-acid.

Lancaster Queen.

REMARKS.—Like Summer Queen, it has a peculiar spicy flavor.—*David Miller*, Pennsylvania Reports.

Lane Sweet. Color, yellow, with red; form, oblong; size, 3; use, table; quality, 3.

LARGE SWEET BOUGH, see *Sweet Bough.*

Lansinburg. Color, green and yellow, with red; form, pippin shape; size, 2; use, kitchen; quality, 3; season, all Winter, Spring, and next Summer.

REMARKS.—"Unworthy of cultivation." — *Trans. Ohio Pom. Society.* Keeps two years. Often exhibited very late in the season. Opinion of the Fruit Committee of the Cincinnati Horticultural Society, "A great keeper, and although not fine for the dessert—far from it—still this apple commands a high price in the Spring. It is sound and firm into Summer." A great keeper, but not worth much when kept, like the Grindstone or American Pippin.

Laquier.

REMARKS.—"Little known in Ohio."—*Trans. Ohio Pom. Society.*

Late Chandler.

Remarks.—Of Cincinnati, Ohio. "Not approved."—*Trans. Ohio Pom. Society.*

Late Strawberry.

Remarks.—"Highly esteemed in New York, where best known."—*Trans. Ohio Pom. Society.*

Leadington, *Monstrous*, or *Green Codlin.*

Remarks.—Very large. Good for cooking.

Lemon Pippin, Kirke's. Color, yellowish green; form, oval; size, 2; use, kitchen; quality, 2; season, September.

Remarks.—Nearly closed at the stalk, and calyx like a lemon. A popular English apple.

Liberty.

Remarks.—Seedling of Delaware county, Ohio. "Good late Winter."—*Trans. Ohio Pom. Society.*

Limber Twig. Color, dull red, and light yellow, striped and splashed with red in sun, rough russet specks, and sometimes a patch of russet; form, round, somewhat conical; size, 2 to 3; use, kitchen; quality, 2; season, keeps till May.

Remarks.—A sound, long keeper, valuable, hardy and very productive. A great market fruit, though rather small. There is said to be a large and small variety. "Valuable for market, being a great bearer, but not a very choice fruit."—*Dr. Warder.*

London Pippin, or *London Sweet.* Color, yellow; form, oblate; size, 2; use, kitchen and table; quality, 2; season, October to January.

Remarks.—Not apt to shrivel like most of the apples

of England. Exhibited at the Cincinnati Horticultural Society. Known in Indiana as the White Winter Pearmain, erroneously. Bears early, productive, and good keeping. Not much valued for the table. Cooks well, and profitable. See Heicke's Winter Sweet. Similar to the Broadwell.

Longville's Kernel, or *Sam's Crab*. Color, brownish red on yellow; form, oval, flattened; size, 3; use, table; quality, 2 to 3; season, August and September.

Long John, or *Kaighn's Spizenburg*.
REMARKS.—"Large, handsome, second-rate." — *Trans. Ohio Pom. Society*.

Long Pearmain, see *Kaighn's Spitzenburg*.

LONG STEM.
REMARKS.—Of Virginia and Eastern Ohio. Resembles Willow Twig.

Lopside, *Grey House*, and *Black Vandevere*, which see.

Loure Queen, or *Lowre Queen*. Color, yellow, striped and splashed with dull red, rough, irregular dots, or specks of dull russet, russet at stem; form, roundish flattened, often angular; size, 2 to 1; use, table; quality, 2; season, October to January.
REMARKS.—Somewhat resembles the Vandevere. Of Morrow county, Ohio. "Large, Winter. Little known." —*Trans. Ohio Pom. Society*.

LOWELL, *or Pound Royal*, *Orange Queen Anne*, *Tallow*. Color, oily, pale yellow; form, oblong ovate; quality, 2; season, August and September.

REMARKS.—Tender, nearly first-rate. A good grower and great bearer. "One of the best Fall apples." — *Trans. Ohio Pom. Society.*

LYMAN'S LARGE SUMMER. Color, pale yellow; form, flattish round; size, 1; use, table; quality, 2.

REMARKS.—Moderate bearer. The tree requires shortening in, like the Peach, to keep up a proper supply of young shoots, as they bear only on the ends.

Lyman's Pumpkin Sweet. Color, green; form, round to conical; size, 1; use, kitchen; quality, 3; season, August to September.

REMARKS.—Rather too large. Often water cored. Not desirable. Was shown before the Cincinnati Horticultural Society as the Reinette d'Espagne.

LYSCOM. Color, red on yellow; form, round; size, 1; use, kitchen and table; quality, 2; season, October to November.

REMARKS.—Flesh, fine grained, mild, sub-acid.

MALE CARLE, or *Mela Carla, etc.* Color, yellowish white, with russet dots, faint orange, or reddish cheek; form, rather flat and globular, with a slight tapering to the eye; size, 1; use, table and cooking; quality, 1 to 2; season, December.

REMARKS.—This fruit is best adapted for the South and South-west, where their seasons of Summer are long. There it is one of the best apples. North it does not answer. In the Middle States it is tolerably good. It is a celebrated Italian apple, not to the fancy of everybody; having, to many, an unpleasant flavor and perfume, which is anything to them but "Rose perfume," as stated by Elliott. It is believed to answer for cooking. Sometimes very beautiful. Mr. R. Clarke, a good and large fruit-grower, likes this apple. He gets a good price for it in market. This apple is generally called medium, but we have them ten inches in circumference: we call that large. They are perfectly sound, now (February 6), and we think they may be called a Winter apple in this climate, although fit to eat in October. Juicy, and fine flavored, but the texture is so fine as to require considerable mastication, until very ripe. Its beauty makes it sell well. It is a full bearer, every year, if well treated. It is of a pale, waxy yellow, with a fine blush on the sunny side, with crimson dots or marbleing; regularly formed; rarely conical.

Maclean's Favorite. Color, yellow; form, roundish; size, 2; use, table and kitchen; quality, 2; season, October to January.

Remarks.—A new English apple. Tree, moderately vigorous, and a good bearer. Not yet tested here. Flavor, like the Newtown Pippin, rich, and crisp in texture. Said to be an excellent fruit where it succeeds.

McLelan. Color, striped red; form, round; size, 2; use, table; quality, 2; season, November to March.

Remarks.—This is described and recommended in the *Horticulturist*, now conducted by R. Pearsall Smith, Philadelphia, the chief horticultural periodical in the Union, as follows: "A beautiful and excellent dessert apple, regular in form; flesh, white, fine grained, very tender and juicy, sprightly, mild, and agreeable, hardly sub-acid."

Maiden's Blush. Color, whitish, blushed, or red cheeked—rather a lemon yellow, with a most beautiful blush; form, flattish; use, table and kitchen; quality, 2 to 1; season, August to October.

REMARKS.—The Maiden's Blush, in this vicinity, is one of the most beautiful (true to its name) and profitable apples cultivated. It is congener to Hawthorndean, or Hawthornden, the best apple in Scotland, but better and more beautiful. A great favorite at the stalls and shops, or stores. Excellent to cook, a prolific and early bearer. Of a peculiar aromatic flavor, and, therefore, not agreeable to some persons—even disagreeable to some. A good market fruit. Rather tartish. "Exceedingly well adapted to the vicinity of Cincinnati."—*F. G. Cary.*

MATSON.

REMARKS.—A large, red streaked, showy, acid, and juicy apple; good bearer; good for cooking; and very marketable; quality, 2.—*Fruits of Missouri, by Thos. Allen, of St. Louis.*

MELT-IN-THE-MOUTH, or *Melting.* Color, yellow ground, overspread with warm yellow red, marbled and dotted with yellow russet; form, roundish, slightly flattened; size, 2; use, table and kitchen; quality, 2; season, November to March.

REMARKS.—From Pennsylvania. Introduced into Ohio. Often exhibited at Pomological Society, at Columbus. Tree, hardy. Fruit, borne mostly on the ends of limbs. Requires a rich soil, abounding in lime. Fruit, small on young trees; improves in size as they grow older. "A good Fall apple."—*Trans. Ohio Pom. Society.*

MAJOR. Color, red; form, round.

REMARKS.—Resembles the Carthouse, or Romanite, but superior. "Very good."—*Pennsylvania Hort. Society.*

Margil, or Neverfail, or Munche's Pippin. Color, orange in the sun, dull yellow in the shade, streaked and mottled with red; form, a little angular, ovate; size, 3; use, table; quality, 2; season, October to January.

REMARKS.—A well flavored, old, English dessert apple, but of too small a size to be popular in the West. Has a high-flavored, aromatic juice.

Menagere. Color, pale yellow, a little red in the sun, sometimes; form, flat; size, 1; use, cooking, only; quality, 3; season, August to January.

REMARKS.—Immensely large. Rather devoid of flavor. Flesh, a little dry.

MARKS. Color, yellowish white; form, round, tapering a little; size, 2; use, dessert; quality, 1.

REMARKS.—"Very good, if not best."—*Pennsylvania Hort. Society.* Flavor delicately perfumed.

MARYLAND. Color, red.

REMARKS.—"Handsome, and good for market."—*Trans. Ohio Pom. Society.*

May.

REMARKS—"Of Myers. Good, long keeper. Little known."—*Trans. Ohio Pom Society.* Outwardly like Pryor's Red. Valued as a keeper only.

MICHAEL HENRY PIPPIN, or *Raritan Sweet.* Color, green, yellowish, lively; form, conical, or roundish oblong, or ovate, narrowing much to the eye, having this peculiarity with some other apples; size, 2 to 3; use, kitchen; quality, 2; season, January.

REMARKS.—This fruit has been confounded with the White Winter Pearmain, which is superior in nearly every respect. "Old, approved Winter."—*Trans. Ohio Pom. Society.* Highly esteemed in some of our Western

States. "A very good little apple in this region. A great bearer, and a good keeper, but not to be compared with White Winter Pearmain, with which it has been so long confounded, it being decidedly sweet. There is, also, a difference in the shape. The tree grows differently, being very upright; and the fruit does not keep as well; the seeds, also, differ."—*T. V. Petticolas.*

MIFFLIN KING. Color, red; form, oblong; size, 3; use, table; quality, 1; season, early in Fall.

REMARKS.—Considered better than the Rambo in Pennsylvania. Very tender, juicy, "first rate."

MILAM, or *Blair.* Color, green and red; form, round; size, 2, sometimes nearly 3 (tree being full); use, table and kitchen; quality, 2 to 1; season, November to February.

REMARKS.—A great bearer. Much grown in Illinois, Ohio and Kentucky, near Cincinnati. Early productive. "Blair, of Chillicothe. Hardy, productive, Winter."—*Trans. Ohio Pom. Society.* Exhibited constantly at the Cincinnati Horticultural Society Rooms. Report of Fruit Committee, "Extensively known throughout the West. In many places it has been propagated by root suckers, by the early settlers. This fruit, though only of second-rate flavor, has so many good qualities, that we desire to present it to the Society. It is a profuse and regular bearer, though of rather small size. It is juicy (unless very late), and tender; may be eaten in the Fall and Winter, and is not difficult to keep until Spring. The country people frequently preserve them in open rail pens, lined and covered with straw. On the contrary, though of a delicate texture, it is deficient in flavor, resembling in its character that of the Westfield Seek-no-further, to which fruit it has many relations. We can not rank it higher than second-rate, but it comes up a *little* beyond this."

MINISTER. Color, red and green; form, roundish, conical; size, 1; use, table; quality, 2; season, September to December.

REMARKS.—"Good, showy."—*Trans. Ohio Pom. Society.* Better for orchard culture in the East than in the West. Needs a deep, sandy loam. Should be carefully picked. Easily rots from bruising. So great a bearer that it often needs thinning. Form rather long ovate. Rather acid. Improved by keeping. Dr. Mosher, of Latonia Springs, thinks well of it. He has paid great attention to fruit.

MONK'S FAVORITE. Color, striped red and yellow; form, globular and flattened; size, 1; use, kitchen; quality, 2; season, till June.

REMARKS.—"Very handsome, promises well."—*Trans. Ohio Pom. Society.* We hope to know more of this fruit.

MONMOUTH PIPPIN, or *Red Cheek Pippin.* Color, greenish yellow, with red cheek; form, round to conical; size, 1 to 2; use, table; season, Winter.

REMARKS.—"Large, good, Winter."—*Trans. Ohio Pom. Society.* A New Jersey apple. Handsome and pretty good.

Monstrous Pippin, see *Gloria Mundi.*

Morgan Sweet.

REMARKS.—"Little known."—*Trans. Ohio Pom. Society.*

MOOR'S SWEETING, *Sweet Pippin, Red Sweet Pippin, of Indiana, and Pound Sweet, of some, West.* Color, yellow striped, and mostly covered with red, white bloom; form, round, flattened; size, 2 to 1; use, kitchen and table; quality, 2; season, early Winter.

REMARKS.—"Exhibited by Mr. Brand, of Champaign

county, Ohio."—*Trans. Ohio Pom. Society.* Good for baking. Flesh firm, rather dry.

MYER'S NONPAREIL. Color, red and yellow; form, flat, roundish; size, 1 to 2; use, table; quality, 1; season, September to December.

REMARKS.—A very handsome apple. "Ohio Nonpareil, of Eastern Ohio, handsome and good."—*Trans. Ohio Pom. Society.* Pronounced by the Cincinnati Horticultural Society, "*The best apple of its season.* Better than the best." Strongly recommended, also, by *Dr. Warder*, President of Society for 1857.

MOUSE APPLE. Color, pale greenish yellow, brownish blush in sun, russet dots; form, roundish conical; size, 1; use, table; quality, 2; season, November to February.

REMARKS.—Sometimes called Moose. Flesh, faintly perfumed. "Very good."—*Trans. Ohio Pom. Society.* A New York Fruit.

MOTHER. Color, yellow, mostly overspread, marbled and striped, with shades of dark red, a few russet dots; form, roundish ovate; size, 1; use, table; quality, 2; season, September to January.

REMARKS.—Common near Indianapolis. Flesh, yellowish, tender, spicy, sub-acid. "Very good."—*Trans. Ohio Pom. Society.*

Murphy. Color, red, striped; form, conical; size, 1; use, hardly known; quality, 3; season, October to February.

REMARKS.—Unworthy of cultivation. "The more I see of this beautiful fruit (from Northern Indiana), the less do I esteem it."—*Dr. J. A. Warder.*

MYER'S APPLE, see *May*.

NEVERFAIL, see *Raule's Janette.*

NEVERSINK. Color, waxen, orange yellow, a few russet dots, striped carmine cheek; form, roundish; size, 1; use, table; quality, 2; season, Fall.

REMARKS.—From Pennsylvania. Flesh, yellowish, flavor a little like the Pine Apple. "Very good."—*National Pom. Society.*

NEWARK KING, or *Hinckman.* Color, red; form, roundish conical; size, 1; use, table and kitchen; quality, 2; season, October to November.

REMARKS.—Not very common. A fine cooking apple. Rather tart, but rich. At any other season it would be considered superior.

NEWARK PIPPIN, or *French Pippin.* Color, green, yellowish when ripe; form, pippin shape; size, 2; use, table and cooking; quality, 1; season, October to January.

REMARKS.—Not sufficiently known or esteemed. Flesh of a very rich yellow. Flavor very fine and rich. "Little known, second-rate, keeps well."—*Trans. Ohio Pom. Society.* We venture to differ from the Ohio Pomological Society, in its being pronounced second-rate. We rank this fruit among the first. It is more tender, and of as high flavor (similar to a good pear), as the highly-prized Newtown Pippin. We obtained this delicious apple from W. Culbertson, on the Licking River, about twenty-six years ago. It delights in a rich, light, vegetable loam. It is a poor and feeble grower. Should be worked up and double worked: yet a strong, growing tree can never be made from one with a weak habit. As a general rule, it is very important for cultivators to select from a nursery good growing kinds. When a good growing variety is once established, although the fruit may not prove good, still there is a good basis to work new fruit upon, when the tree is vigorous. Good growers, with bad fruit, can be altered to bear good; there is a strong base to work upon; but weak growers can not be changed to produce very good fruit, or much of it. Here will be a loss of time and money. How beneficial it is to obtain good stocks for every purpose.

NEWTOWN SPITZENBURG. See *Ox Eye and Joe Berry.*

NEWTOWN PIPPIN, YELLOW. Color, greenish yellow when ripe, with a red cheek; form, round, often one-sided when large, russety in and a little round the stalk; size, 1; use, chiefly table, but good in kitchen, although a little crisp; quality, 1; season, March.

REMARKS.—Highly approved in suitable localities. Apt to spot in soils not adapted to it. From Long Island. It requires a rich, limestone clay soil, or a warm, sandy, rich loam, well dressed with lime and bone dust. It is distinct from the Green Newtown Pippin, described next; and the rich limestone soil of Ohio, etc., suits both. The Yellow Newtown has a higher flavor than the Green. On sandy soil, not generally good; apt to speck.

NEWTOWN PIPPIN, GREEN. Color, dull green when first gathered, when ripe a yellowish green, with small russet dots, with occasional blotches of the same, and on alluvial soils, South, patches of dark green mold; use, table; quality, 1; considered, generally, superior to the Yellow Newtown; season, January to May.

REMARKS.—Bears alternate years. Superior. Exhibited often at the Cincinnati Horticultural Society Rooms. Fruit Committee report it "a good bearer, of high flavor, and excellent when ripe (in March). Not the most digestible before it is perfectly matured. Good for the kitchen at any time during the Winter. Near Cincinnati it is subject to be injured and disfigured with hard, black spots, generally considered of vegetable origin, or fungus, but they do not induce decay. "Newtown Pippin is often inferior. If the stem of this tree is trimmed up to full standard hight, the bark becomes very dry and rough, and the top seems to starve, even in a rich soil."—*American Pom. Society, received from reports from Pennsylvania.* Both the Yellow and the Green Newtown have russet marks at the stalk.

NEWTOWN SPITZENBURG, or *Flushing.* Color, red on yellow; form, round, flattened, often one-sided; size, 1; use, table; quality, 1; season, November to February.

REMARKS.—"Ox Eye of Cincinnati."—*Trans. Ohio Pom. Society.* One of the best table fruits we have for December. Grown North, it keeps longer, and is neater and smaller, more regular and less wormy, and less liable to fall than with us. It is highly esteemed in the West. How Downing came to call this old familiar fruit Vandevere, is a wonder. It came from Long Island. Sometimes called Joe Berry, in Kentucky, as well as Ox Eye, at Cincinnati, and in the West.

Nickajack Apple.

REMARKS.—A Southern fruit, and very good.

NONSUCH (see Red Canada), *Richfield*, or *Canada Red.* Color, red on yellow; form, roundish conical; size, 2; use, table; quality, 2; season, December to February.

REMARKS.—Superior. Exhibited at Horticultural Society Rooms by J. E. Mottier. Declared Baldwin, erroneously. Much grown North.

6

Nonsuch, Old.

Remarks.—Not esteemed. Grown close to the gate of the late Dr. Smith, near Cincinnati.

Norfolk Beaufin.

Remarks.—Rarely seen in this country; and, we presume, not desirable. It is a large fruit. Color, dull red on greenish. Good for drying only.

NONPAREIL, see *Ross's Nonpareil*, which is a highly flavored, spicy russet, not known nor cultivated to any extent. An apple with this name (Nonpareil), is grown in Illinois, large, sub-conical, irregular, or ribbed; red striped, on yellow ground, with a lively bloom; flesh, yellowish, very delicate, and good; as beautiful as Northern Spy, which it resembles, as, also, like Scolloped Gilliflower.

Northern Spy. Color, striped red on light yellow ground, with streaks of carmine red, and when first gathered, covered with a fine bloom; form, sub-conical, sometimes ribbed; size, 1; use, table; quality, 2; season, November to January, in this latitude becoming an early Winter variety.

Remarks.—There is a great diversity of opinion about this apple with us. It does well, as near us as Dayton. It is about the dividing line. There is a different soil there—more gravely. Sold at Boston for six dollars per barrel, and sometimes one dollar per dozen, by retail. It keeps, there, till May. Juicy, very crisp and fresh. "Of doubtful value in Ohio."—*Trans. Ohio Pom. Society.* It is probably much over estimated by James H. Watts. The perfume is extraordinary, and the flavor sometimes very good. A great grower, and not early in fruiting. We fear its great excellence as a keeper may not be realized here (Cincinnati). We must wait a little. This variety, however, like the Gate, or Waxen and Baldwin, etc., is ripened too soon in this latitude to be a first-rate Winter fruit, losing much of the excellent flavor possessed by them where found in perfection. "If ever the trees of this fruit come into bearing with us, we will be able to say something more about them. We have had them grafted seven years without fruit yet."—*T. V. Petticolas.* The Northern Spy may probably become a good apple here. It has succeeded well at Dayton; Mr. Sayers having informed the author that he has seen very fine specimens of that good fruit there.

Old English Codlin, see *Codlin.*

Orange Apple.
REMARKS.—"Not deemed valuable."—*Trans. Ohio Pom. Society.*

ORANGE SWEETING, *Golden Sweet*, or *Trenton Early.*
REMARKS.—Flourishes well in all soils. Yields fine crops of fair fruit. Tree of medium size, branches straggling. A good fruit.

ORNDORF. Color, lemon yellow, rich red blush in sun, stripes and blotches of red; form, roundish, slightly angular; size, 2; use, table; quality, 1 to 2; season, September to November.
REMARKS.—"Of Putnam, Ohio. Little known."—*Trans. Ohio Pom. Society.* Flesh, yellowish, juicy, crisp, tender, sub-acid. Really a good apple.

ORTLEY PIPPIN, with some twenty synonymes; among them, best known, are *White Bellflower, Hollow Core Pippin, Detroit*, and *Golden Pippin.* Color, pale yellowish white at the North, a richer yellow more South, with some specks of dark red; form, oblong oval, sometimes roundish conical; size, large, or No. 1 on rich soils; use, table; quality, 1; season, November to March.

REMARKS.—"One of the most agreeable and digestible of all apples, with a mild, sub-acid, abundant juice, without any remarkable or high flavor."—*Trans. Ohio Pom. Society.* Excellent in strong soils. This is one of our prime favorites. Thrifty, productive, delicate, certainly not highly flavored, but most easy of digestion. Its size is much increased in the West. Originally from New Jersey.

Oslin.

REMARKS.—Not yet known.

OX APPLE, or *Ox Eye*, and *Joe Berry;* see *Newtown Spitzenburg*.

REMARKS.—A delicious apple, as before described, but drops worse than any apple we have. By the time they are ripe they are all gone, not only on old, but young trees. We can not recommend it on that account, but it may do better in some soils. A neighbor, whose trees are all dead from age, etc., did not remark that fault.

PEACH POND SWEET. Color, striped red and yellow; form, oblong; size, 2; use, table; quality, 1; season, early September to last of October.

REMARKS.—A great bearer. A favorite dessert apple of all who taste it. From the orchard of R. Buchanan, Esq., Clifton, Cincinnati.

Pennock's Red Winter, or *Phœnix*, by some, the same as the Large Romanite, of Kentucky. Color, red; form, roundish flattened, almost invariably one-sided; size, 1;

use, chiefly kitchen, though light for the stomach, in eating; quality, 3; season, December and January.

REMARKS.—Exhibited often, and rather late in the season. Fruit Committee of Cincinnati Horticultural Society consider it third-rate, and condemn it as a table apple, although extensively propagated and planted, being very vigorous, and a large tree that bears abundant crops. Texture, tender, light, and very easily digested, but rather harsh and a little astringent, with considerable sweetness. It is rather inclined to rot and spot inside and externally, and is best adapted for cooking.

PECK'S PLEASANT, or *Watts' Apple.* Color, clear yellow, blush on sunny side; form, round, slightly flattened, indistinct furrow on one side; size, 2 to 1; use, table; quality, 1; season, November to February.

REMARKS.—On sandy soils, of a firmer texture than on clays. "Valuable on sandy soils."—*Trans. Ohio Pom. Society.* This fruit deserves more attention than it has yet received from our pomologists. It is fine wherever we have seen it. Sometimes this apple keeps till April.

PHILLIP'S SEEDLING, or *Sweeting.* Color, yellow ground, nearly entirely covered and mottled with red; form, roundish conical, a little flattened; size, 2 to 1; use, table; quality, 2; season, November and December.

REMARKS. — Requires a strong clay or heavy soil. Growth, vigorous and upright. Native of Central Ohio. "Large, handsome, good, Winter."—*Trans. Ohio Pom. Soc.*

Pine.

REMARKS. — "Of Morrow county, Ohio. Resembles Seek-no-further."—*Trans. Ohio Pom. Society.*

PINE APPLE RUSSET, or *Hardingham's Russet.* Color,

greenish yellow, covered with thin russet; form, roundish ovate; size, 2 to 1; use, table; quality, 2; season, September and October.

REMARKS.—Flesh, yellowish white, juicy, crisp, spicy, sub-acid. Of foreign origin.

PINK SWEETING. Color, red and pink stripes; form, fine; size, 3 to 4; use, table; quality, 2.

REMARKS.—Bears immense crops. Has a spicy flavor. Excellent for stock feeding.

PITTSBURGH FAVORITE. Color, greenish white and yellow, at maturity; form, flat; size, 1; use, table; quality, 1; season, November to February.

REMARKS.—Not very productive. Of a high, pleasant sub-acid flavor. As fine nearly, if not quite, as Rhode Island Greening.

Polly Bright.

REMARKS.—"Little known. Approved by some."—*Trans. Ohio Pom. Society.* Of Pennsylvania and Eastern Ohio. Resembles Maiden's Blush, and like it sharp and acid. Season, September and October.

POMME GRISE. Color, yellowish gray or russet; form, roundish, rather flat; size, 3; use, table; quality, 2; season, October to February.

REMARKS.—Best adapted for gardens. Better for the North than here. A good bearer, of fine flavor, and keeps well. Perhaps the very best of the russets. Shown at the Cincinnati Horticultural Society Rooms, by R. Reilly, from Butler county, Ohio. Very fine specimens.

POMME DE NEIGE, or *Snow Apple*, or *Fameuse.* Color, greenish yellow, mostly overspread with pale and dark rich red; form, roundish, somewhat conical; size, 2; use, table; quality, 2 in Southern Ohio, 1 in Canada and in the North; season, October and November.

REMARKS.—"Canadian origin. Very dark color. Good." —*Trans. Ohio Pom. Society.* Sometimes streaked in the North. A beautiful Northern fruit, good size, perfect form, smooth skin, white flesh, juicy. Prolific and early bearer. Flavor not high nor rich. Cooks well. It ripens too soon here (Cincinnati), and will not be profitable in this neighborhood.

POMME ROYAL, *White Seek-no-further*, *Flushing Seek-no-further.* Color, yellowish green, small brown dots; form, roundish oblong, conical, uneven, or waved surface; size, 2 to 1; use, table; quality, 1; season, October to January.

REMARKS.—From Long Island. Flesh, fine grained, very juicy, tender, sub-acid. The Pomme Royal, or Dyer, is not the Woodstock, as some authors assert. We have seen them both at one of our State Fairs. The Woodstock is a coarse apple, only fit for cooking. The Pomme Royal is one of the very best dessert apples.

Pound, or *Monstrous Pippin.* Color, yellowish green; form, roundish, oblong; size, 1; use, table and kitchen; quality, 3; season, September to December.

REMARKS.—Flesh, coarse, poor, unworthy.

Pound Royal.

REMARKS.—"Of Dr. Barker. Dyer."—*Trans. Ohio Pom. Society.*

PORTER. Color, yellow; form, conical; size, 2; use, table and kitchen; quality, 2; season, August.

REMARKS.—"A popular Eastern Fall apple. Little known in Ohio."—*Trans. Ohio Pom. Society.* Celebrated in Massachusetts. Not admired here generally. Tart, not high flavored. Cooks well. "Exceedingly well adapted to the vicinity of Cincinnati. Requires lime and phosphate."—*F. G. Cary, Pres't Farmers' College.* This apple deserves a place in every orchard, for its beauty alone; but independent of that, it is well flavored. Nothing can exceed in the apple line, a basket of these in beauty; in proof of which we will merely mention the fact of Mr. Petticolas getting one dollar and fifty cents per bushel for them, when other apples were only bringing sixty cents.

Priestley. Color, green bluish purple; form, conical; size, 2 to 1; use, table and kitchen; quality, 3; season, Winter till April.

REMARKS.—"A long keeper, of second-rate quality."—*Trans. Ohio Pom. Society.* Condemned.

PRINCE'S EARLY HARVEST. See *Early Harvest.*

Prolific Beauty.

REMARKS.—"Of the Putnam list. Worthless."—*Trans. Ohio Pom. Society.*

PRYOR'S RED. Color, pale yellow, warm red and russet, having the appearance of being smoked sometimes; form, regular, roundish; size, 2 to 1; use, table and kitchen; quality, 1 (for market chiefly); season, November to April.

REMARKS.—Not an early bearer, but very hardy. Keeps well when grown South. The richest of russets. Rather shy in bearing. Exhibited rather seldom before the Cincinnati Horticultural Society. Judgment of Fruit Committee, "In fine condition, firm and fresh, late in Winter." This fruit is deserving of high commendation for its many good qualities. It varies much in its appearance, being sometimes green russeted, then dull orange russet, without a trace of red, and again deeply red or striped, and sometimes almost black with depth of color. There may be different varieties, but all have richness of flavor, and the form that contains the greatest amount of material within a given compass, having a very small cavity and basin. A native of Virginia, it does well in the Middle States. Though rather long coming into bearing, may be set down as a No. 1 apple.

Pumpkin Russet.
REMARKS.—Miserable.

Pumpkin Sweet.
REMARKS. — "Very large. Good for stock." — *Trans. Ohio Pom. Society.* Fruit Committee of Cincinnati Horticultural Society, "Third-rate."

Pantneyite.
REMARKS.—"A Virginia apple of little value."—*Trans. Ohio Pom. Society.*

PUTNAM RUSSET. Color, russet; form, round to flat, often one-sided; size, 1; use, cooking, and then rather mealy and coarse; quality, 2; season, December.

REMARKS.—Overrated. Bears well; thrifty. Falls badly. A very good keeper. Flesh, coarse and tart. Cooks pretty well. It is an old-fashioned apple. A good keeper. Tart, without flavor, and valuable in its bearing qualities. Sometimes large and fair, at other times small and knotty. It is thin skinned.

QUEEN, see *Fall Queen.*

RAMBO, or *Seek-no-further of Pennsylvania.* Color, a

yellowish white; form, roundish, a little flat; size, 2; use, table; quality, 1; season, October to February.

REMARKS.—Succeeds in all soils, and situations. Hardly any superior as a Fall apple. Best, however, in limestone soils. "Generally known and approved."—*Trans. Ohio Pom. Society.* "The Rambo, the Smokehouse, and Fall Pippin are preferred to all others of the season, for the table (so far as varieties have been proved here)."—*American Pom. Society, Washington.* This variety is found wherever Pennsylvania Germans settle. Some varieties exist, keeping better than others, which ripen too soon. With us (Cincinnati), it is generally large, and is a Fall fruit, which becomes very dry and mealy in November. Still, the Rambo is not overrated. If not allowed to remain too long on the tree, they remain juicy and crisp to the last, but they vary very much on the same tree. Its greatest fault is overbearing on alternate years.

RAMBOULETTE.

REMARKS.—"Of Judge Wood. Rambo."—*Trans. Ohio Pom. Society.*

Rambour Franc.

REMARKS.—Poor.

RAMSDALE'S SWEETING. Color, red striped; form, roundish conical; size, 1; use, table and kitchen; quality, 1; season, October.

REMARKS.—This is Red Pumpkin Sweeting. A good table and baking fruit. Dr. Mosher, of Latonia Springs, values it highly. (He is a good pomologist.) One of the best and handsomest sweet apples cultivated in this country, being tender.

RARITAN SWEET. Color, whitish; form, round;

size, 2; use, kitchen and stock; quality, 1; season, December.

REMARKS.—We consider it valuable. "Of Morrow Co., Ohio. Good, Winter."—*Trans. Ohio Pom. Society.*

RAULE'S JANET, *Genneting, Neverfail, etc., etc.* Color, light pale yellowish green, stained with dull red, with small russet dots, sometimes high colored; form, roundish conical, flat at stem; size, 2; use, both table and kitchen; quality, 1; season, late Spring—often best in April.

REMARKS.—Tardy in coming out in Spring. Generally escapes Spring frosts. Very long keeper. Good for Kentucky, Tennessee, Arkansas, Missouri, Illinois, Indiana, and Southern Ohio. Flesh, juicy, sweet, lively, and very pleasant; tender, when ripe, in late Winter or Spring. "Exceedingly well adapted to the locality of Cincinnati." —*F. G. Cary.* Cincinnati Horticultural Society's Fruit Committee consider it generally "a prime favorite." The tree is very prolific, and should be thinned, and well fed, to produce choice specimens. Its late blooming has caused it to be named "Neverfail," as it thus often escapes frosts that have killed the blossoms of others, as the Bellflower, etc. It is called by the Scriptural name of Rock Rimmon, in the Scioto Valley, on account of its sure bearing, and long keeping qualities. "Recommended unanimously."—*Trans. Ohio Pom. Society.* The more red stripes there are in it, the better the fruit. The more green, the worse. Its greatest fault is, a too great tendency to bear. This is only to be remedied well by pruning, both inside and out, or thinning and shortening in. It is a great matter to adapt the pruning to the habit of the tree. Some varieties want a great deal, and some but little.

RED ASHMORE.

REMARKS.—"Handsome, good."—*Trans. Ohio Pom. Soc.*

Red Bellflower, Striped Bellflower, etc. Color, greenish yellow, covered and striped with red; form, oblong conical; size, 1; use, table, but poor; quality, 3; season, October to December.

REMARKS.—Foreign. Unworthy.

Red Calville.

REMARKS.—"Little known, and less value."—*Trans. Ohio Pom. Society.*

RED CANADA, *Old Nonsuch, Richfield Nonsuch.* Color, rich, clear yellow ground, red to the sun, striped light and dark red, with gray spots; form, roundish conical, flattened at stem end; size, 2; use, table; quality, 1; season, November to April.

REMARKS.—"Excellent, North and South."—*Trans. Ohio Pom. Society.* Requires the best of soils. Always fair and regular in shape. Excellent, either for orchard or garden. A native of Massachusetts.

RED FAVORITE. Color, red; form, flat; season, Fall.

REMARKS.—Pleasant, juicy, sub-acid.

Red Winter Calville.

REMARKS.—Unworthy, here.

RED, or BLACK GILLIFLOWER. Color, dark red; form, conical, long; size, 1; use, sale, from its size and show; quality, 3; season, January to March.

REMARKS.—The Black Gilliflower is a standard market fruit, being productive, and a good keeper, fine quality, rather dry, but high flavored. Not much cultivated in this locality (Cincinnati).

RED GILLIFLOWER. Color, light red, striped; form,

conical, scolloped, or ribbed; size, 1; use, table; quality, 2; season, November to December.

REMARKS.—An early Winter fruit, of fair quality. Not equal, however, to many others of its season. Little known here (Cincinnati).

Red Streak, or Early Red Streak. Color, red striped; form round; size, 2; use, kitchen; season, July.

REMARKS.—Exhibited by many at the Cincinnati Horticultural Society's Rooms. Fruit Committee consider it poor. "Unworthy."—*Dr. Warder.* Pennsylvania Red Streak is synonymous with Hay's Red Winter. Red Streak is also applied to some other fruits. An early, coarse, tart, rich-looking apple, with heavy bloom. Good for cooking. Sells remarkably well. Two trees, planted twelve years, brought Mr. Petticolas twenty-six dollars, or thirteen dollars to the tree. It bears full crops in alternate years, and half a crop the other.

RED ASTRACHAN. Color, deep red; form, flat, roundish; size, 2; use, kitchen; quality, 1; season, August.

REMARKS.—"Valuable for early market."—*Trans. Ohio Pom. Society.* This will prove, when better known, a valuable fruit. Tart, but good when cooked with sugar. A productive and early bearer. A very beautiful, early apple. In 1855 they came in before Prince's Harvest, and sold well. They bear on alternate years only, and are long coming into bearing. The tree is quite ornamental.

Red and Long Pearmain, same as, or synonymous with Kaighn's Spitzenburg, also Long John, and Scarlet Pearmain.

REMARKS.—A cooking apple. Bears every other year. An inferior apple, not worth cultivating.

Red Gilliflower.

REMARKS.—We have a very fine, large apple by this name, very tender and of good flavor, bright red in color, but a shy bearer on young trees. Mr. Mears says, they bear well when the trees get old. Ours have been planted nine years, and have not borne a dozen to the tree, yet.

Red and Green Sweet.

REMARKS.—"Deemed unworthy of cultivation."—*Trans. Ohio Pom. Society.*

Red Ingestrie. Color, dark red; form, round; size, 3; use, kitchen; quality, 2; season, September and October.

REMARKS.—This little apple will never prove popular here. "Foreign. Small, rich; little known in Ohio."—*Trans. Ohio Pom. Society.*

RED GROVE.

REMARKS.—Regarded by the National Pomological Society as "very good."

RED QUARRENDEN, *Devonshire Quarrenden, or Sack Apple.* Color, deep, clear red, with specks of russet green; form, roundish flattened; size, 2; use, table, and cooking; quality, 2; season, August.

REMARKS.—"A handsome and popular, Eastern and foreign fruit, little known in Ohio."—*Trans. Ohio Pom. Society.*

Red Russet, Golden Pearmain, Dutch Pearmain, or Ruckman's Pearmain—pronounced by Ohio Pomological Society, *Goble Russet.*

RED WINTER PEARMAIN.

REMARKS.—One of our best Winter Pearmains. Ten

years ago it was the best apple in this region, but since then it has become scabby almost every year, until this last year (1856), when it resumed its old character, and we hope it will continue to maintain it.

REINETTE, BLANCHE D'ESPAGNE, or *White Spanish Reinette.* Color, yellowish green; form, roundish oblong, sometimes ribbed; size, 1; use, cooking; quality, 2; season, October to January.

REMARKS.—A variety of the Fall Pippin—probably the parent of that fruit. Sharp sub-acid. Only fit for cooking.

Reinette Pearmain.

REMARKS.—"Little known, and of little value."—*Trans. Ohio Pom. Society.*

Reinette Triumphante. Color, pale yellow; form, roundish oblong; size, 1; use, table; quality, 2; season, November and December.

REMARKS.—Victorious Reinette. German. Rich, tart. Not yet known here. Found in Mr. Rehfuss's garden.

REPUBLICAN PIPPIN. Color, striped red; form, roundish, flattened; size, 1; use, table; quality, 2; season, Fall.

REMARKS.—Flesh, yellowish white, tender; flavor, pleasant, peculiar (slightly sub-acid), resembles sometimes that of walnut. "Seems to bear poorly, and the first fruits, at least, are not fair. Uncertain, and liable to speck and rot."—*Trans. American Pom. Society.* Described by Dr. Brinckle, in *Downing's Horticulturist.*

RHODE ISLAND GREENING. Color, greenish yellow, when ripe; form, round, flattened; size, 1; use, table, and cooking; quality, in this vicinity, hardly second-rate sometimes, though when sound and healthy always 1; season, November to January.

Remarks.—One of the best cooking apples, and quite rich for table when ripe. Except with Mr. W. Orange, who grows it on a sloping ground, with a north-east aspect, it is too large and spongy in these parts, becoming russety. It here generally falls and decays badly. Very different in Rhode Island, and elsewhere, to what it is here. There a very fine and valuable apple. Requires a very rich soil, the lime and phosphates, which is probably the reason it does not do well with us (Cincinnati), being subject to bitter rot. "Highly approved, but liable to rot and speck in most localities, West."—*Trans. Ohio Pom. Society.* Flesh, rich yellow. Grows well in the North in a thin sand. It varies much in different localities. With Mr. Petticolas it has done very well so far, bearing good crops of fine apples. Rather tart for the dessert, but excellent for cooking. None of the Eastern apples are so compact, here, as there, but they are generally larger and more spongy. It sells well.

Ribston Pippin, Glory of York, Travers, and Formosa, or Beautiful Pippin. Color, greenish yellow, russet near the stem, dull red in the sun; size, 2; use, table and kitchen; season, October to January.

Remarks.—The finest apple in England. The flesh, there, rich, firm, yellow, aromatic sub-acid. Valuable in Northern regions, a failure in the South, or Central. "English, excellent in some localities, chiefly North."—*Trans. Ohio Pom. Society.* "Does not equal, by any means, and at a long distance, its European and Northern character, in the West."—*Dr. J. A. Warder* (one of our best pomologists). Doubtful if we have this apple genuine. There is a great difference of opinion on that subject. The specimens we have seen here, are so different from the largest and most beautiful, and richest apple in England, that we can not but hesitate in thinking them genuine.

Robinson.

Remarks.—"Little known, and of little value."—*Trans. Ohio Pom. Society.*

ROCK RIMMON, see *Raule's Janet.*

ROME BEAUTY. Color, bright red, yellow ground; form, roundish; size, 1; use, table; quality, 1 to 2; season, November to February.

Remarks.—Fruit hangs on the tree late; keeps well through Winter; commands a high price from its fine size and great beauty. Improves by remaining on the tree late. This fruit, in a rich soil, grows sometimes gigantically large. Immensely large ones have been exhibited before the Cincinnati Horticultural Society. Not highly flavored, but a good fruit. Very attractive for the stalls and markets. Flesh, yellow, tender and juicy, with slight sub-acid, and sweet, agreeable flavor. "A seedling of Southern Ohio, highly approved for market and orchard culture."—*Trans. Ohio Pom. Society.* Sells often at four dollars per barrel. It is one of the greatest bearers, of large showy apples, in the whole catalogue. It bears every year. It keeps well and sells well; is very profitable, but only a good common flavored apple.

Roman Stem. Color, red on yellow; form, round obovate; size, 2; use, table; quality, 1; season, November and December.

Remarks.—"A good Winter apple. Keeps well."—*Trans. Ohio Pom. Society.* Valuable for Pennsylvania, and elsewhere, and fine in the prairies of Illinois. Productive.

ROMANITE, see, also, *Gilpin* and *Carthouse.* "Gilpin. Well known. Very long keeper."—*Trans. Ohio Pom. Society.* Exhibited often before the Cincinnati Horticultural Society. Fruit Committee decide it only third-rate for the table, having an earthy taste. The trees are very early and prolific bearers; the fruit is remarkably regular, and sound, and may be easily preserved until May or June, and will command about four dollars per barrel at Cincinnati. It has been found to produce a very rich cider.

ROBEY'S SEEDLING.

REMARKS.—Pronounced "very good" by the National Pomological Society.

ROSS'S NONPAREIL. Color, dull red russeted; form, round; size, 3; use, table; quality, 1; season, September to November.

REMARKS.—Productive. Highly flavored and valuable.

ROXBURY RUSSET, *Putnam Russet, Marietta Russet, Boston Russet, Belpre Russet,* and *Sylvan Russet.* Color, dull green, yellow russet, occasionally faint blush on sunny side; form, round, rather flat; size, 1 to 2; use, table and cooking; quality, 3 for table, 1 to 2 for cooking; season, October to December.

REMARKS.—Rather dry when kept. Hardy and very productive. Falls rather badly. A thick skin. "Putnam Russet, same as Roxbury."—*Trans. Ohio Pom. Society.*

Rule's Summer Sweeting. Color, yellow; form, oblong; size, 2; use, poor; quality, 3; season, July and August.

REMARKS.—Poor. Unworthy; but can be made available for stock, etc.—as all sweet apples are more nourishing to hogs, cattle, etc., than sour or sub-acid ones.

SCOLLOP GILLIFLOWER. See *Gilliflower Scolloped.*

REMARKS.—"Handsome. Little value."—*Trans. Ohio Pom. Society.*

SCARLET PEARMAIN. Color, red; form, oblong, conical; size, 1; use, table and kitchen; quality, 2 to 3; season, November and December.

REMARKS.—"Little known in Ohio."—*Trans. Ohio Pom. Society.* This fruit is much cultivated, though not without faults. Much admired by Mr. A. H. Ernst. Sometimes very handsome, when large. Has many small specimens, which are always poor. See Kaighn's Spitzenburg, identical.

Scarlet Sweeting.

REMARKS.—"Of Morgan county, Ohio. Handsome, little known."—*Trans. Ohio Pom. Society.*

Scarlet Nonpareil, or New Scarlet Nonpareil. Color, deep red on yellowish green; form, roundish; size, 2; use, kitchen and table; quality, 3; season, October to January.

REMARKS.—Flesh, firm, acid. Foreign.

SEEK-NO-FURTHER, WHITE. Color, white; form, round, flattened a little; size, 1; use, table; quality, 1; season, August and September.

REMARKS.—Exhibited by T. V. Petticolas (a great fruit cultivator, twelve miles from Cincinnati), often; he calls it White Seek-no-further, and "very good," which is concurred in by Horticultural Society's Fruit Committee. "Westfield."—*Trans. Ohio Pom. Society.* We disagree. Not identical with White Seek-no-further.

SEEVER'S SEEDLING, of Coshocton county, Ohio.

REMARKS.—"Handsome, good."—*Trans. Ohio Pom. Soc.*

SILVER RUSSET.

REMARKS.—The finest flavored russet known, when you can get them, for they almost invariably rot before ripe.

SINE QUA NON. Color, greenish yellow; form, roundish ovate; size, 2; use, table; quality, 2; season, August.

REMARKS.—Flesh, juicy, sprightly, sub-acid, and excellent flavor. Slow, poor grower. Good bearer. Our rich Western soils undoubtedly give both fruits and vegetables a larger and more tender, or spongy growth. And so this fruit has become larger, and the tree a better grower here. "Of Long Island. Poor bearer, and of little value."—*Trans. Ohio Pom. Society.* This conflicts with our experience near Cincinnati; but localities differ.

Small Black.

REMARKS.—"Of T. V. Petticolas. Not approved."—*Trans. Ohio Pom. Society.* Only fit for cider.

SMITH'S CIDER. Color, pale bright red and yellow, sometimes deep red, nearly all red, with white specks; form, roundish flattened, slightly oblong sometimes; size, 2 to 1 in favorable situations and soils; use, table and cooking; quality, 2 to 1; season, November to February.

REMARKS.—Fine, good grower, with a spreading habit. This fruit is generally considered deserving of a better name; cider inferring an apple mostly suitable for that purpose only. It is a fruit which has, in our view, considerable character, and its peculiar and aromatic flavor is liked by many. It is mostly fair and glossy, and attracts much attention in the stalls. Some persons think it has but little flavor, and that unpleasant. The palate will not admit of much disputation, being rather an arbitrary organ. White flesh, juicy. "Handsome and good Winter."—*Trans. Ohio Pom. Society.* We propose that it be called Smith's Apple, *formerly Cider*. But "give a dog a bad name," etc. This apple, take it all in all, is the most profitable grown, bearing heavy crops, every year, of large, splendid looking, perfect apples, scarcely a worm to be found in them. They present the grandest sight imaginable in the orchard about picking time, the tall trees being covered from top to bottom, with large crimson, brilliant apples. In quality, they are about like Rome Beauty, but a little more brisk and aromatic, keeping until late in the Spring, and always selling well. This day (5th February), we see them selling at twenty-five cents for four, but they are of large size.

Smith's Summer.

REMARKS.—Best for drying. Wood very thrifty; top regular. A good annual bearer. Fruit, large, oblate, regular, sweet, juicy.

SMOKE HOUSE. Color, striped red; form, rather flat; size, 2 to 1; use, table; quality, 2 to 1; season, September and October.

REMARKS.—Very productive. Flesh, yellowish white, crisp, and juicy. Flavor, agreeable, with a delicate aroma. Exhibited before the Cincinnati Horticultural Society frequently. Much grown in Pennsylvania. Fruit Committee consider that it can be long preserved in good order, though generally an early Winter fruit. Much grown also in the North-West. It is a Pennsylvania variety. A seedling from the Vandevere, but better than that apple. It is described by Mr. Brinckle, in the *Horticulturist*, Philadelphia, and much esteemed by him, an excellent pomologist. We think it will be a profitable apple. It shows symptoms of being a great bearer. It is new here yet.

Sops of Wine. Color, crimson, darker in the sun; form,

flattish, conical; size, 3; use, table; quality, 3; season, August.

REMARKS.—Crispy, juicy, tolerably pleasant, sub-acid. Beautiful, but neither excellent nor profitable.

Spence's Early, or Sweet June, Summer Sweet of Ohio, etc. Color, greenish yellow, with greenish white dots; form, roundish; size, 2; use, table; quality, 2; season, July.

REMARKS.—Flesh, yellowish white, tender, juicy, sweet. Exhibited before the Horticultural Society by R. Buchanan, in 1855. The Fruit Committee consider it productive, but apt to crack in the ripening. Good for stock.

Spitzenburg, Kaighn's.

REMARKS.—A coarse, but very showy apple. A tolerable bearer. Large. Sells well.

Springer's Seedling.

REMARKS.—"Long keeper; not first-rate."—*Trans. Ohio Pom. Society.*

STURMER PIPPIN. Color, yellowish green, and brownish red; form, short, conical; size, 2; use, table; season, until Midsummer.

REMARKS.—Tree healthy, and a good bearer, highly flavored, and brisk.

SPICE SWEETING. Color, yellowish; form, roundish; size, 1; use, table; quality, 2; season, July and August.

REMARKS.—"Handsome, high flavored."—*Trans. Ohio Pom. Society.*

SPICE PIPPIN, *Ortley, or White Bellflower;* see *Ortley, etc.*

REMARKS.—Often exhibited before the Society. Found excellent, and to succeed well in this vicinity.

STANLEY'S EARLY.

REMARKS.—"Resembles Early Pennock. Second-rate." —*Trans. Ohio Pom. Society.*

STROAT, or *Straat*. Color, yellowish green; form, roundish conical; size, 1; use, table; quality, 2; season, September to November.

REMARKS.—The Dutch name for Street. Flesh, yellow, tender, brisk sub-acid. Not much grown in Ohio. A pleasant, late Summer fruit, but tart. At Dr. Wm. Smith's, near Cincinnati. An excellent apple. Very much like White Seek-no-further, but a shy bearer.

STRIPED BELLFLOWER. See Red Bellflower.

Summer Calville. Color, greenish yellow, reddish on one side near the stem, cavity russet; size, large; form, oblate.

REMARKS.—Wood, thrifty; top, not very regular; subject to blight.—*Fruits of Missouri*, by T. Allen of St. Louis.

SUMMER SWEET. See Spence's Early, or Sweet June.

REMARKS.—The earliest of sweet apples—July and August. Has no superior in its season.

SUMMER CHEESE, *Fall Cheese, or Gloucester Cheese*. Color, greenish yellow, flush of red in the sun; form, roundish; size, 2 to 1; quality, 2; season, August and September.

REMARKS.—"Of Eastern Ohio. Good, early sweet."—*Trans. Ohio Pom. Society.*

SUMMER ROSE. Color, glossy yellow, with red stripes; form, roundish, somewhat flattened; size, 3; use, table; quality, 1; season, July and August.

REMARKS.—Tree, feeble in growth. Not valuable for

market, but for dessert. "Fine, but slow grower."—*Trans. Ohio Pom. Society.* "Excellently well adapted to the vicinity of Cincinnati."—*F. G. Cary,* Farmers' College, Ohio. This is thought by the American Pomological Society the best apple between Yellow Harvest (or Prince's Yellow Harvest), and Summer Queen, and, with them, it proves an excellent bearer. Exhibited by F. G. Cary, July, 1855. Fruit Committee pronounced it, "gradually ripening, crisp, brisk, cooking well, and fine for dessert. Should be planted by every one who has room for a single tree."

SUMMER QUEEN. Color, yellow, red stripes; form, conical; size, 2; use, table and kitchen; quality, 1 to 2; season, July and August.

REMARKS.—"Good, especially for market."—*Trans. Ohio Pom. Society.* One of the richest Summer cooking apples. Too tart for dessert. Exhibited by R. Buchanan, Esq., and others, July, 1855. Fruit Committee regard this apple as a "highly flavored, acid fruit, very fine for cooking."

SUMMER PEARMAIN. See Autumn Pearmain, Sigler's Red, etc.

REMARKS.—Excellently well adapted to the vicinity of Cincinnati.

Shipley, Green. Color, red and russety; form, oblong; size, 2; use, baking; quality, 3; season, very long.

REMARKS.—Pennsylvania apple. Very sour, and long keeper. "Sheepnose, of Mr. Petticolas. What is it?"—*Trans. Ohio Pom. Society.* The author considers it very like the Newark, or French Pippin, except the flesh is not so high a yellow, nor the flavor so rich; the form is the same, conical, pippin-shaped; the skin is also less green before it is ripe, when it is of a light yellow. Mr. Mears, of Fruit Committee, brought the same apple before the Horticultural Society, under the name of Sheepnose. Sheepnose is properly a synonyme of the American Golden Russet.

SIBERIAN CRAB.

REMARKS.—There are many varieties, as Red, Large Red, Yellow, Purple, Striped, Transparent, Oblong, Double White, Fragrant, Cherry, Showy, Astrachan, Currant. All used only for Preserving, or grown for ornament. The Large Red Siberian Crab is in possession of the author. It is about twice the size of the foregoing, roundish ovate, with a large and prominent calyx, and rich and bright red and yellow skin. It grows in the thickest clusters, all along the branches, and produces a most beautiful and brilliant appearance at a small distance off, like red and yellow cherries. It is a superior kind of this fruit, and by no means common. It makes rich, firm, and beautiful Preserves, when gathered before it is too ripe. The stalks should remain on. Boughs of this fruit have excited the admiration of all, when exhibited before the Cincinnati Horticultural Society.

SIGLER'S RED, *Autumn Pearmain, Royal Pearmain, Summer Pearmain, etc.* Color, brownish yellow and green, red blended with yellow in sun, small brown specks; form, oblong conical; size, 2; use, table and kitchen; quality, 2; season, August and September.

REMARKS.—Foreign. Tree, slender, slow growth, irregular. Flesh, pale yellow, crisp, nearly "best."

Summer Golden Pippin. Color, yellow; form, roundish oblong; size, 3; use, table; quality, 3; season, August.

REMARKS.—Flesh, whitish, firm, sweet. Unworthy of cultivation.

8

St. Lawrence.—Color, striped dark red, green ground; form, roundish, slightly oblate, sometimes conical obtuse; size, 2; use table and market; quality, 2; season, September and October.

Remarks.—A decorative fruit, to which class it properly belongs. Of Canadian origin. "Not to be highly recommended."—*Western Horticulturist.*

Surprize.

Remarks.—A fancy sort, with a pink flesh. Unworthy.

Sugar Loaf Pippin, or Sugar Loaf Greening. Color, greenish yellow; form, oblong, conical; size, 1; use, table; quality, 3; season, July and August.

Remarks.—Hardly worth notice, although "good," or third-rate.

SWAAR (in Dutch, heavy). Color, dull green, when gathered, becomes, toward Spring, of a brilliant lemon color, with brown specks; form, round, flattened, slightly ribbed; size, 1 to 2; use, table; quality, 1; season, February to March.

Remarks.—Requires a rich soil. "Old Winter apple. Good in some localities."—*Trans. Ohio Pom. Society.* A rich and valuable sub-acid apple for Spring. Hardly known with us. Much grown North.

Sweet Seedling.

Remarks.—From Indiana. A premium recommended for it. Flavor, good.

Sweet Bellflower, *Butter* of some. Color, lemon yellow, slight blush in sun, numerous light and dark specks; form, globular, flattened at base, slightly ribbed; size, 1; use, table; quality, 2; season, October and November.

Remarks. — This description is from A. H. Ernst. "Esteemed in some parts of Ohio."—*Trans. Ohio Pom. Society.* Flesh, whitish yellow, breaking, juicy, sub-acid. Large, handsome, and good, but a shy bearer.

Sweet Bellflower, of Wyandot county. Color, light yellow, dark, cloudy flakes, and yellow specks; form, globular, slightly conical; size, 1; use, table; quality, 2; season, November and December.

Sweet Gilliflower. Very like the above, if not identical.

Sweet Cann.

Remarks.—From New Jersey. Winter, fine cooking. Exhibited by W. S. Chapman, at the Horticultural Society Rooms, November, 1855.

Sweet Pearmain.

Remarks.—"Little known. Good long keeper."—*Trans. Ohio Pom. Society.*

Sweet Romanite, *Sweet Nonsuch, Orange Sweet* (erroneously). Color, greenish yellow ground, striped with bright red, and has a fine bloom; form, roundish, flattened, and regular; size, 2; use, table; quality, 1; season, November to March.

Remarks.—From W. B. Lipsey, Morrow county, Ohio. In Illinois it is grown as the Sweet Nonsuch. Flesh, greenish yellow, firm, crisp, juicy, superior to Ramsdell's or Danver's Sweeting, or Winter Sweet.

Sweet Pippin. "Of Dayton. Good. Late Fall."—*Trans. Ohio Pom. Society.*

Sweet Bough. Color, greenish, pale yellow when ripe.

REMARKS.—"Large Yellow Bough. Early, good, generally approved.—*Trans. Ohio Pom. Society.* Profitable for market. Common in Cincinnati market, and popular.

Sweet London Winter, or London Winter Sweet.

REMARKS.—One of our finest Winter sweet apples, uniformly fair, and regularly formed. A great bearer on alternate years. Pale straw color, oblate. Sells well, and the very best for apple butter.

Sweet Lyman's Pumpkin, or Lyman's Pumpkin Sweet.

REMARKS.—Does remarkably well here, bearing large, fine-looking apples. Excellent for stock, and the tree bears very young.

SOUR OR TART BOUGH. "Resembling Early Harvest; not so good."—*Trans. Ohio Pom. Society.* Profitable also for market. Much seen in Cincinnati market, and popular.

SWEET RUSSET, *Pumpkin Russet, York Russet*, and *Flint Russet.* Color, yellowish green, thinly russeted; form, roundish; size, 1; use, cooking, for apple butter; quality, 2; season, August to October.

REMARKS.—An excellent fruit for cooking in cider. We have a butter apple here that is valuable in its season for cider and apple butter. Both that and the Sweet Russet are fair and great bearers.

TART BOUGH. See Sour Bough. The best of the Boughs.

TALPEHOCKEN. See Fallawater.

REMARKS.—"Large. Second-rate."—*Trans. Ohio Pom. Society.*

TALLOW APPLE, or LOWEL. Color, oily, pale yellow.

REMARKS.—Very much like the Porter; more acid. Bears very young.

Tewksbury Winter Blush, see Fink. Color, yellow, red cheek; form, rather flat; size, 3; use, table; quality, 3; season, February, to July or August.

REMARKS.—Pleasant, but not high flavored. Remarkable for freshness after long keeping. Vigorous and productive. Cultivated in the Middle and Western States. A good baking and stock apple. "Valuable for long keeping."—*Trans. Ohio Pom. Society.*

TALMAN'S SWEETING. Color, whitish; form, roundish conical; size, 2; use, table and stock; quality, 1.

REMARKS.—"First-rate, especially for baking in Winter."—*Trans. Ohio Pom. Society.* Good for baking and stock, particularly with us. It has proved, generally, a very inferior fruit—not at all to be compared with the Broadwell, which is the best keeping sweet apple for this climate.

TETOFSKY. Color, a yellow ground, handsomely striped with red; and, like most Russian apples, covered with a whitish bloom, under which is a shiny skin; form, roundish oblong, sometimes nearly round; size, 2; use, table; quality, 2; season, August, sometimes July.

REMARKS.—This apple is not much known yet in this country. What is known of it is favorable. It is said to be valuable for cooking and marketing.

TITUS PIPPIN.

REMARKS.—Not generally known; resembles Newark, or French Pippin, and Ortley. A fine looking and pretty good large apple. Looks a little like Yellow Bellflower. A fine, upright tree, bearing heavy crops. Profitable.

TOWNSEND APPLE. Color, pale yellow; form, roundish, and usually flattened; size, 2; use, table; quality, 1; season, August to September.

REMARKS.—Bloom like the Astrachan. "One of the most delicious late Summer and early Autumn apples."—*Horticulturist.* "Not highly recommended."—*Trans. Ohio Pom. Society.*

TRENTON EARLY.

REMARKS.—"Resembles Gold Sweeting."—*Trans. Ohio Pom. Society.*

Twenty Ounce. Color, green, striped or blushed; form, round conical; size, 1; use, kitchen; quality, 2; season, October.

REMARKS.—A drying kind; popular in Western New York. Unworthy of introduction from its season. "Cayuga Red Streak; second-rate."—*Trans. Ohio Pom. Society.*

TYRONE. Color, greenish yellow, red stripes and white dots; form, pippin shaped; size, 1; use, table and kitchen; quality, 2; season, Winter to Spring.

REMARKS.—Brought from Ireland by Wm. Culbertson, on the Licking River; a pioneer fruit-raiser here.

Victuals and Drink, Big Sweet, or Pompey.

REMARKS.—Unworthy.

Vandevere Yellow, or Vandevere Pippin. Color, red and yellowish green, striped gray; form, round, flattened a good deal; size, 1; use, kitchen; quality, 2 to 3; season, September to December.

REMARKS.—Coarse and acid; excellent for cooking and drying. Much cultivated in Western Pennsylvania, Ohio, and Indiana; weighs very heavy. This is different from the gray Vandevere. Apt to speck and rot. Large tree, thrifty and productive. Cooks well with sugar, but coarse and tart when raw.

Virginia Greening. Color, green; form, round, flat; size, 1; use, kitchen; quality, 3; season, very late in Spring, or beginning of Summer.

REMARKS.—Unworthy. "Long keeper, but valueless." —*Trans. Ohio Pom. Society*. Exhibited at the Cincinnati Horticultural Society's Rooms very late in the season. The Fruit Committee "can not recommend this variety for the table, yet it keeps well, and will command a good price in the market in the Spring." It is a passable apple next Summer—very different from the Grindstone, or American Pippin. This used to be a great favorite in this neighborhood (now fortunately times have changed), but it has lost of late its character of a good keeper. It rots badly on the trees. We consider it a very poor apple, and not to be recommended.

WAGENER. Color, shaded, and indistinctly striped with pale red, and a full, deep red in the sun, on warm, yellow ground, often streaked with russet; form, oblate, obscurely ribbed; size, 1; use, table; season, ripens through the Winter.

REMARKS.—Much admired in New York; flesh yellowish, very fine grained, tender, compact, mild, sub-acid, very aromatic; excellent. Should be freely trimmed, to produce large, fine fruit in abundance.

WATSON'S DUMPLING. Color, yellowish green, faintly striped; form, nearly round, regular; size, 1; use, for cooking; quality, 2; season, early Winter, or late Fall, in Ohio.

REMARKS.—English. Indifferent, poor.

Ward.

REMARKS.—"Of Champaign County, Ohio; very large and showy; little known."—*Trans. Ohio Pom. Society.*

WAXEN APPLE. See Belmont.

WELL'S APPLE. Color, bright yellowish, green and red; form, roundish conical, slightly oblate; size, 2 to 1; use, fine baking; season, Winter.

REMARKS.—Tree very productive, Central Ohio. "English Red Streak, Striped Rhode Island Greening, Domine? Approved in Eastern Ohio."—*Trans. Ohio Pom. Society.*

WHITE BELLFLOWER, *Detroit*, *Ortley*, etc. See Bellflower, White.

REMARKS.—Very superior, but very subject to scabbiness. Bears heavily every other year, and were it not for scab would be far before Yellow Bellflower.

WILLOW TWIG. Color, greenish yellow, striped and mottled faintly with dull red; form, roundish, slightly conical; size, 1; use, table; quality, 2; season, Winter.

REMARKS.—A long keeper. Cultivated with us much now as a market apple. "Profitable Winter. South Western Ohio."—*Trans. Ohio Pom. Society.*

WINE, synonymous with *Hay's Apple, or Hay's Winter.* Color, obscurely striped and mottled with red, on yellow ground; size, 2 to 1; use, table; quality, 2; season, Fall.

REMARKS.—Flavor, rich, sub-acid. Flesh, yellowish white, very handsome and regular; not equal to many other sorts, however. There are several apples improperly called by this name. "Profitable and good Winter." —*Trans. Ohio Pom. Society.* It has a pleasant, vinous flavor, whence its name.

WINESAP. Color, dark red, lively; size, 2; use, kitchen and dessert; quality, 2 to 1; season, November to February.

REMARKS.—Hardly ever fails to bear. No. 2 in quality and No. 1 in profit. A valuable, second-rate apple. A productive and early bearer. "Exceedingly well adapted to the vicinity of Cincinnati."—*F. G. Cary.* Very good grower; hardly a word to be said against it. Valuable for cultivation. One of the good little apples, or rather medium, pleasant eating all the Winter, good cooking, and a good bearer. Rather too small to command a great price in market.

WESTERN SPY.

REMARKS.—"Of Central Ohio. Handsome and good Winter."—*Trans. Ohio Pom. Society.*

WESTFIELD SEEK-NO-FURTHER.

REMARKS.—This apple is entirely distinct from the White Fall Seek-no-further, the Westfield having red and russet at one end. More like Pryor's Red, being a medium Winter apple. The White never has any red or russet, but is remarkable when the apple is ripe (a pale straw color), for a green tinge around the stem, and is gone in November.

WHITE DETROIT, see *White Bellflower and Ortley.*

White Gilliflower.

REMARKS.—"Of Mr. Benedict; not approved."—*Trans. Ohio Pom. Society.*

WHITE CODLIN. Color, white; form, oblong; size, 3; use, table and kitchen; quality, 2; season, November and December.

REMARKS.—Pennsylvania apple, originated in Maryland. Fall and early Winter. Very rich, acid juice, and very pleasant.

WHITE BELLFLOWER, or *Detroit*, see *Ortley, etc.* Color, light yellow; form, generally conical, a little flat at base; size, 1; use, dessert; quality, 1; season, October to March.

REMARKS.—Skin, thin, smooth and oily to the touch. Flesh, tender and sprightly. Identical with *Ortley, or Woolman's Long, White Detroit, Hollow Core Pippin, Ohio Favorite, etc., etc.* —"Excellently well adapted to the vicinity of Cincinnati."—*F. G. Cary*, a practical fruit cultivator. One of the most hardy and profitable apples in the West, and much seen on the stalls.

WHITE SPICE. Form, regular round, somewhat flat; size, 1; use, table; quality, 2; season, August and September.

REMARKS.—A Pennsylvania apple. "Good for home use, but especially for market."—*Pennsylvania Reports, from David Miller, Jr.*

WHITE PIPPIN, see *Canada Pippin* (although identity is very doubtful), and *Canada Reinette.* Color, greenish white; form, round; size, 1; use, kitchen and dessert; quality, 1; season, Winter.

REMARKS.—Succeeds well in the vicinity of Cincinnati. "Good and profitable Winter."—*Trans. Ohio Pom. Society.* Exhibited by M. McWilliams (one of the Fruit Committee of the Cincinnati Horticultural Society, with whom it is, as with others, deservedly a very great favorite), and others. Pronounced by the Fruit Committee, at large, of great value. By some it is considered to lack high character and flavor, but if it is not of the very highest, we think it makes very close approaches to them. Excellent for cooking, also, and productive.

White Winter Calville.

REMARKS.—Not admired. Grown somewhat in Canada and in the North.

WHITE WINTER PEARMAIN. Color, narrow and broken stripes of dull red or greenish yellow; form, oblong-ovate conical, or verging to a point, though ends somewhat flattened; size, 2; use, dessert; quality, 1; season, as its name implies.

REMARKS.—This apple is of the highest excellence, and early engaged the attention of the Fruit Committee of the Cincinnati Horticultural Society. They have thus described it:—"Medium, conical, basin shallow, often plaited, stem short, skin smooth, greenish yellow when ripe,blushed when exposed; flesh, firm, breaking, juicy, very sweet, pleasant, lively, sub-acid; seeds of a peculiarly light brown color, in a moderate cavity." This very prolific and delicious variety is a good keeper, and highly valuable. How could it have been confounded with Michael Henry Pippin so long? We consider this the best flavored dessert apple of its season (April and May), retaining its juicy flavor and crispness to the last. They were fine on the 10th of last June (1856). A great bearer on alternate years; a little subject to scab. The wood is remarkably soft, stems thick, and when loaded with fruit hangs to the ground; just the reverse of the Michael Henry Pippin, which is never pendulous.

WHITE JUNEATING.

REMARKS.—"Bracken, and Carolina, of Southern Ohio." —*Trans. Ohio Pom. Society.*

WHITE RAMBO. Color, white; form, roundish flattened; size, 1 to 2; use, table and kitchen; quality, 2 to 1; season, Fall.

REMARKS.—By some preferred to the famous Rambo. It is larger and whiter. "Of Morrow county. Approved where known."—*Trans. Ohio Pom. Society.* I. Morris sells more trees of this than of any other variety. Not generally distributed. Has the same thick skin and dots, as the common Rambo.

White Vandevere.

REMARKS.—"What is it?"—*Trans. Ohio Pom. Society.*

White Astrachan. Color, white, faint streak of red; form, roundish conical; size, 2; use, table; quality, 3; season, July and August.

REMARKS.—Unworthy—poor.

WHITE SEEK-NO-FURTHER.

REMARKS.—"Resembles Westfield, except in color."—*Trans. Ohio Pom. Society*

Williams' Favorite.

REMARKS.—Has not met the expectations of its planters. Sometimes very beautiful.

WILLOW LEAF.

REMARKS.—Willow Leaf, and Willow Twig, Mr. Mears says, are distinct. Willow Leaf is rather a poor apple; its keeping qualities being its only recommendation. We, however, think these two apples are the same.

WINTER PEARMAIN, WHITE. See, also, White Winter Pearmain. Color, greenish yellow; form, conical;

size, 2; use, table; quality, 1; season, January to March.

REMARKS.—"Good keeper; second-rate."—*Trans. Ohio Pom. Society.* We consider it nearer No. 1, near Cincinnati. Exhibited late in the season before the Horticultural Society, and is fast growing to be a favorite. It is sound and good till Spring. Prolific, and a good keeper; nearly first-rate in quality; a safe-keeping variety, and a juicy and good apple in January and February. Often confounded with the Michael Henry Pippin, from which it is not easily distinguished by the taste; seeds, pale brown, while those of the Michael Henry, are nearly black. The shape is also more conical — hence its synonyme, Sheepnose. Michael Henry often becomes cottony and tasteless. White Winter Pearmain is crisp and juicy. See Michael Henry.

Winter Cheese.

REMARKS.—Of Eastern Ohio. Good as a long keeper.

WINTER QUEEN. Color, deep crimson in the sun, a lively, pale red in the shade; form, conical, base broad; size, 1; use, table; quality, 2; season, early Winter.

REMARKS.—It is a large, fine, juicy fruit.

WINTER GREENING.

REMARKS.—"Of Zanesville. Resembles Rhode Island Greening."—*Trans. Ohio Pom. Society.*

Winter Pennock. See Pennock.

REMARKS.—"Coarse, liable to bitter rot."—*Trans. Ohio Pom. Society.*

Winter Sweet Pippin.

REMARKS.—"Little known."—*Trans. Ohio Pom. Society.*

WING SWEETING. Color, light and dark red, indistinctly striped on dark yellow; form, round, flattened; size, 2; use, table; quality, 2; season, Winter.

REMARKS.—Productive.

WINTER SWEET PARADISE.

REMARKS.—A rich, sweet apple, of good, sprightly but not high flavor; light in weight; a very productive and excellent, good sized orchard fruit, and of fine appearance. "Excellent for table, and baking."—*Trans. Ohio Pom. Society.* A valuable sweet apple, grown much about Columbus by Pennsylvanians. Successfully cultivated in different parts of Ohio. Generally approved.

Wonderlich's Spice.

REMARKS.—"Resembles American Golden Russet."—*Trans. Ohio Pom. Society.*

Woodstock.

REMARKS.—A coarse apple, only fit for cooking. It has been confounded with the Dyer, one of the very best dessert apples.

Wormsley Pippin. Color, greenish yellow; form, roundish; size, 2; use, cooking; quality, 3 to 4; season, Fall.

REMARKS.—Unworthy.

Yellow Ingestrie. Color, clear, rich yellow; size, small.

REMARKS.—Of little value in this country.

YELLOW NEWTOWN PIPPIN. See also Newtown Pippin, Yellow, and Newtown Pippin, Green. Color, greenish yellow; form, round flattened, often one-sided; size, 1; use, table and kitchen; quality, 1; season, March.

REMARKS.—The American apple, in England, at the

tables of Royalty and the Nobility, and the rich who can afford the luxury. There is only one apple there that comes up to this — the Ribston Pippin — and that has hardly as high a flavor. "Well adapted to the locality of Cincinnati, with a rich soil."—*F. G. Cary.* Rich limestone lands best. Pell grows it largely for shipment in New York. The Yellow is the commonest kind in market, and is larger than the Green. Apt to be one-sided. Richer color, more juicy and sprightly, and less tender fleshed than the Green Newtown. Rather indigestible before Spring, but of the highest character for flavor. There is great confusion about these two kinds in nurseries.

YELLOW BELLFLOWER. See Bellflower. Color, as its name denotes; form, oblong, conical; size, 1; use, table and kitchen; quality, 1; season, December.

REMARKS.—"Generally approved, rather acid."—*Trans. Ohio Pom. Society.* A favorite fruit, of peculiar flavor. Very tart before ripe. A wonderful change when ripe. Apt to be frosted in the blossom. "Exceedingly well adapted to the locality of Cincinnati."—*F. G. Cary.*

York Russet.

REMARKS.—"Sweet Russet. Large, sweet, poor."—*Trans. Ohio Pom. Society.*

Zane's Seedling.

REMARKS.—"Resembles Gilpin or Romanite."—*Trans. Ohio Pom. Society.*

Zane's Greening

REMARKS.—"Not generally approved."—*Trans. Ohio Pom. Society.*

ADDENDA OF SEEDLINGS.

Buchanan's Pippin, *or James's.* Color, dull greenish red; form, round; size, 1; use, table; quality, 2; season, March to May.

Remarks.—This pleasant seedling was named after a young son of Mr. Robert Buchanan, and found by him in his father's fine orchard. It is named after him, as a melancholy memento. Basin open, planted moderately deep. Tree with a round, upright head. A great bearer, and vigorous.

Seedling, *from A. A. Mullett.* Color, yellow, with red streaks; form, evenly round, rather flattened; size, 2; quality, 2; season, fine eating beginning of October.

Remarks.—Sub-acid, and good for cooking.

Fink's Seedling.

Remarks.—A remarkably long keeping variety, of good quality. Raised by John Clarke, of Somerset. Specimens grown one year, well preserved, were shown along with those of the next season. Size, 2, very smooth and round; color, dull green, becoming yellow at maturity, with a dash of bronze red on the sunny side; flesh, white, tender, juicy, of mild, sub-acid flavor. The tree of fair growth, and very productive.

Longworth's Sweet. Color, yellowish green, with spots of smoky russet; form, roundish, conical; size, medium; use, table; quality, 2; season, Winter.

Remarks.—Flavor, very pleasant, but rather dry in February. A very good sweet apple, particularly in December.

LIST OF APPLES,

FOR THE WESTERN STATES.

The following list contains a Catalogue of the most popular variety of Apples recommended by various Pomological Societies of the United States for the Western States:

BALDWIN. Ohio, Missouri, Illinois.

ROXBURY RUSSET. Michigan, Ohio, Missouri, Indiana, Illinois.

RHODE ISLAND GREENING. Michigan, Iowa, Ohio, Missouri, Illinois.

SWAAR. Ohio, Illinois, Michigan.

ESOPUS SPITZENBURG. Ohio, Missouri, Illinois, Michigan.

EARLY HARVEST. Virginia, Ohio, Missouri, Indiana, Illinois, Michigan, Iowa.

SWEET BOUGH. Illinois, Virginia, Missouri, Indiana, Ohio.

SUMMER ROSE. Ohio, Missouri, Illinois.

FALL PIPPIN. Michigan, Virginia, Ohio, Missouri, Illinois.

BELMONT. Michigan, Ohio.

GOLDEN SWEET. Missouri.

RED ASTRACHAN. Iowa, Ohio, Missouri, Illinois.

JONATHAN. Ohio, Missouri.

EARLY STRAWBERRY. Ohio.

DANVER'S WINTER SWEET. Ohio.

AMERICAN SUMMER PEARMAIN, Illinois.

MAIDEN'S BLUSH. Ohio, Missouri, Indiana, Illinois.

PORTER. Ohio, Missouri.

GRAVENSTEIN. Ohio.

VANDEVERE. Missouri, Indiana, Illinois.

YELLOW BELLFLOWER. Michigan, Iowa, Virginia, Ohio, Missouri, Illinois.

FAMEUSE. Illinois.

NEWTOWN PIPPIN. Michigan, Iowa, Virginia, Ohio, Missouri, Indiana, Illinois.

RAMBO. Michigan, Iowa, Ohio, Missouri, Indiana, Illinois.

SMOKEHOUSE. Virginia, Indiana.

FALLAWALDEN. Ohio.

GOLDEN RUSSET. Ohio, Illinois.

WINESAP. Ohio, Illinois.

WHITE BELLFLOWER. Missouri, Illinois.

HOLLAND PIPPIN. Michigan, Missouri, Indiana.

RAULE'S JANET. Iowa, Virginia, Illinois.

LADY APPLE. Ohio, Missouri.

LIST OF APPLES,

ADAPTED TO ORCHARD AND GARDEN CULTURE IN THE OHIO VALLEY.

	Names.	Summer.	Fall.	Winter
1.	American Summer Pearmain	1	0	0
2.	American Golden Russet	0	0	1
3.	Bellflower (Yellow)	0	0	2
4.	Belmont	0	0	3
5.	Bevan's Favorite	2	0	0
6.	Benoni	3	0	0
7.	Black's Annett	4	0	0
8.	Bohannon	5	0	0
9.	Broadwell	0	0	4
10.	Cooper	0	1	0
11.	Carthouse	0	0	5
12.	Domine	0	0	6
13.	Dutch Mignonne	0	0	7
14.	Early Strawberry	6	0	0
15.	Early Chandler	7	0	0
16.	Early Pennock	8	0	0
17.	Findley	9	0	0
18.	Fall Pippin	0	2	0

Names.	Summer.	Fall.	Winter.
19. Fameuse (Snow Apple)	0	0	8
20. Fort Miami	0	0	9
21. Gravenstein	0	3	0
22. Harvest (Yellow H., Early H.)	10	0	0
23. Jersey Sweet	0	4	0
24. Jonathan	0	0	10
25. Keswick Codlin	11	0	0
26. Large Sweet Bough	12	0	0
27. Lady Apple (Pomme d'Api)	0	0	10
28. Limber Twig	0	0	11
29. Maiden's Blush	0	5	32
30. Michael Henry Pippin	0	0	10
31. Milam	0	0	14
32. Newtown Spitzenburg	0	0	15
33. Ortley (White Bellflower)	0	0	16
34. Pennsylvania Red Streak	0	6	0
35. Phillips' Sweeting	0	0	17
36. Pryor's Red	0	0	18
37. Rambo	0	7	0
38. Raules' Janet	0	0	19
39. Red Astrachan	13	0	0
40. Rhode Island Greening	0	0	20
41. Rome Beauty	0	0	21
42. Roxbury Russet	0	0	22
43. Summer Queen	14	0	0
44. Summer Rose	15	0	0
45. Smokehouse	0	8	0
46. Smith's Cider	0	9	0
47. Vandevere	0	0	23
48. White June (Juneating, etc.)	16	0	0
49. Williams' Favorite	17	0	0
50. Willow Twig	0	0	24
51. Wine Apple	0	0	25
52. Yellow Newtown Pippin	0	0	26

The list, as thus extended, embraces fifty-two varieties of apples, which competent authorities recommend for general cultivation in the Ohio valley. Of these, seventeen are Summer apples, and twenty-six Winter apples.

1. "Decidedly the best apple of its season."—*Trans. Ky. Hort. Society.* "Highly approved."—*Trans. Ohio Pom. Society.* "Requires a deep, warm soil, well supplied with lime and potash, when it succeeds admirably in all sections.—*Elliott.* "Bears early and abundantly; one of the best in all parts of the country."—*Barry.*

2. "First rate."—*Trans. Ohio Pom. Society.* "Flesh remarkably tender; juicy, almost buttery, delicate, sprightly."—*Elliott.*

3. "Generally approved; rather acid."—*Trans. Ohio Pom. Society.* "Tender, juicy, sprightly, sub-acid."—*Elliott.* "Crisp, juicy, pretty acid, and rich. Very productive, succeeds well throughout all portions of the country."—*Barry.*

4. "Generally approved, especially in Northern Ohio, but in Southern part of the State somewhat given to rotting upon the tree."—*Trans. Ohio Pom. Society.* "Flesh very tender, juicy, sprightly, sub-acid. On all high, warm, or limestone soils, does finely."—*Elliott.* "Succeeds well in New York and Northern Ohio, but is variable at Cincinnati and further South. Flesh, sub-acid, juicy, fine."—*Barry.*

5. "Of little value."—*Trans. Ohio Pom. Society.* "Flesh, fine, tough, sub-acid."—*Elliott.* "A New Jersey apple, where it is esteemed as one of the best of its season, sub-acid and good."—*Barry.*

6. "Handsome, early, and good."—*Trans. Ohio Pom. Society.* "Flesh, yellow, tender, crisp, juicy, vinous, very good."—*Elliott.* "Tender, juicy, and rich; a good bearer."—*Barry.*

7. "Recommended by Young and Byram. A very

superior apple, well known, and deservedly popular."—*Trans. Ky. Hort. Society.*

8. "Fine Southern apple."—*Trans. Ohio Pom. Society.* "Flesh, yellowish white, tender, slightly aromatic, sub-acid."—*Elliott.* "Very delicious, high flavored, very tender, sprightly, and fine."—*Barry.*

9. "Sweet, approved where known."—*Trans. Ohio Pom. Society.* "Flesh white, fine-grained, sweet, juicy."—*Elliott.* "Tender, sweet, and excellent."—*Barry.*

10. "Highly recommended by many."—*Trans. Ohio Pom. Society.* "Flesh, yellowish, not fine-grained, crisp, juicy, very good."—*Elliott.* "Tender, juicy, and agreeable."—*Barry.*

11. "Good keeping qualities, flesh yellowish, firm, juicy."—*Elliott.* "Sub-acid and agreeable. Largely cultivated in some parts of the South, where it is esteemed for its productiveness and good keeping qualities."—*Barry.* [The Ohio Pomological Society makes Gilpin and Romanite synonymes. Elliott's synonymes are Carthouse and Romanite of the West. Barry's, Gilpin, and Red Romanite.] The Ohio Pomological Society calls the Gilpin "small, good keeper, second-rate."

12. "A pleasant Winter apple."—*Bateham.* "Flesh, white, tender, juicy, very good."—*Elliott.* "Sub-acid, juicy, and high-flavored. Resembles Rambo, and, like it, succeeds well, West and South."—*Barry.*

13. "Proved valuable wherever grown; very fine in Southern Ohio. Flesh, whitish, firm at first, becoming tender when well matured, sub-acid, aromatic."—*Elliott.* "Beautiful and excellent apple; fine flavor, good bearer." —*Barry.*

14. "Fine and early."—*Trans. Ohio Pom. Society.* "Productive and successful in all localities. Flesh, yellowish white, tinged with red, sub-acid, sprightly, tender."—

Elliott. "Tender, almost melting, with a mild flavor; good bearer."—*Barry.*

15. "Handsome, high-flavored, acid."—*Trans. Ohio Pom. Society.* "Good quality; extensively cultivated in some parts of Ohio, where it succeeds well."—*Barry.*

16. "Large, handsome, second-rate."—*Trans. Ohio Pom. Society.* "Tree, thrifty, hardy; early, prolific bearer. Fruit, rather below second-rate quality. Flesh, yellowish white, juicy, sub-acid."—*Elliott.* "Large, handsome, and showy."—*Barry.*

17. Recommended by Young and Byram. Local, and a fine Kentucky fruit. Believed by Col. Anderson, of Meade county, to be identical with the "Horse Apple." Called by some old citizens, the "Runnels," and "Fort Runnels Apple." Trees, vigorous, and great bearers. Fruit, quite large, ripening the latter part of July; yellowish green color; flesh, white, mild, somewhat sprightly, very juicy, and palatable. Grows very large on the gravely loam undulations of the Peewee Valley. Good for either dessert or cooking. Much liked by stock. Lasts till September. Preferred, by the writer, to the Early Harvest, as more juicy and sprightly.

18. "Large, handsome, and good."—*Trans. Ohio Pom. Society.* "Universally succeeds well. Flesh, yellowish white, tender, sub-acid, aromatic."—*Elliott.* "Tender, rich, and delicious; a fine bearer. Fruit, esteemed everywhere."—*Barry.*

19. A great favorite in Bourbon county, Kentucky, and highly approved by Mr. Bedford. Elliott says: "Without being a fruit of high character, it is just so good, that, taken with its production of regular, handsome fruit, it can not be dispensed with. Flesh, remarkably white, tender, juicy, with a slight perfume." "Tender, and delicious."—*Barry.*

20. "New, and productive. Flesh, yellowish white, fine-grained, tender, mild, and sub-acid."—*Elliott.* "Said to be rich, and high-flavored; a good keeper."—*Barry.*

21. "A good fruit."—*Trans. Ky. Hort. Society.* "Handsome, and good."—*Trans. Ohio Pom. Society.* "Indispensable to every collection; succeeds finely in all soils; annually productive; fruit, always fair, fit for cooking in August. Flesh, yellowish, crisp, tender, sub-acid, with a peculiar aromatic taste."—*Elliott.* "Tree, very productive, and fruit, of first quality."—*Barry.*

22. "Well known, and everywhere approved."—*Trans. Ohio Pom. Society.* "Universally esteemed; requires a soil well supplied with lime and potash. Flesh, white, tender, juicy, crisp, sub-acid."—*Elliott.* "Rich, sub-acid. Tree, a good bearer."—*Barry.*

23. "First-rate for table."—*Trans. Ohio Pom. Society.* "Succeeds in all localities. Abundant bearer in all soils. Warm, sandy soils give more character to the flesh, and a closer texture. Flesh, white, fine-grained, tender, juicy, sweet."—*Elliott.* "A good bearer; succeeds well in all parts of the country; much esteemed everywhere for dessert and cooking."—*Barry.*

24. "One of the handsomest and best apples."—*Trans. Ohio Pom. Society.* "Very productive, but needs rich, strong soil. Flesh, yellowish white, tender, juicy, slightly acid until fully matured, then sub-acid, and sprightly."—*Elliott.* "Very productive. Flesh, tender, juicy, and rich, with much of the Spitzenburg character."—*Barry.*

25. "A popular Summer cooking apple."—*Trans. Ohio Pom. Society.* "Very productive; valuable for cooking; one of the best for Western soils. Flesh, greenish white, tender, acid."—*Elliott.* "Bears abundantly quite young; acid; excellent for cooking from July to October."—*Barry.*

26. "Early, good, generally approved."—*Trans. Ohio Pom. Society.* "Tree, a moderate, annual bearer, succeeding

in all good soils not wet. Valued as a dessert fruit. Flesh, white, tender, crisp, sprightly, sweet."—*Elliott.* "Abundant bearer, sweet, rich flavored."—*Barry.*

For the descriptions of the remainder, we refer the reader to the several names in their proper places in this work.

FRUITS OF OHIO.

Statement of R. BUCHANAN, A. H. ERNST, *and* J. A. WARDER, *of Cincinnati, Hamilton county, Ohio, as reported to the American Pomological Society at their annual meeting held at the City of Boston in September,* 1854.

"The climate and soil of our State are so varied, and the fruit in culture so numerous, that a report to embrace catalogues to suit each locality would be too voluminous. From Cleveland, on Lake Erie, in the Northeast, to Cincinnati, on the Ohio, in the Southwest, a distance of two hundred and fifty miles, there is a difference of near three degrees in latitude, and a great diversity of soil. It is, therefore, difficult to fix a uniform standard of excellence in fruits for the whole State.

"Loam and clay, intermixed with lime and sand, are the principal components of our soil, often underlaid by a substratum of gravel, and the greater portion of our State is well adapted to the culture of most of the fruits grown in the Middle States.

"The present report will be confined to the Southwestern and Central parts of our State. The Northeastern section having been embraced in previous reports.

APPLES.

"The average bearing of apple trees, with us, is four out of five years. Many varieties which are highly esteemed further North do not suit the climate and warm limestone soils of Southern Ohio. The 'Rhode Island

Greening,' for instance, ripens and casts its fruit so early here as to become a Fall apple, and but few are gathered from the trees for Winter. The 'Alexander,' with us, is an early Fall apple, and the far-famed 'Esopus Spitzenburg' is here a shy bearer, and an unprofitable variety to cultivate. Even the 'Baldwin' and the 'Roxbury Russet' mature too early, and do not keep so well as when cultivated further North and in cooler soils. The 'Belmont,' a favorite apple in Northern and Eastern Ohio, with us is subject to crack open, and rot upon the tree in some seasons.

"With ordinary care and culture, the apple thrives well in all parts of our State, and, with the exception of the grape, is the most certain bearer of any of our fruits. The following list comprises the most favorite varieties cultivated in this section.

"SUMMER VARIETIES.—Benoni, Bohanon, Drap d'Or, Carolina Sweet, Early Bough, Early Harvest, Gravenstein, Maiden's Blush, Red Astrachan, Strawberry, Summer Rose, Summer Pearmain, Summer Queen.

"FALL VARIETIES.—Alexander, Belmont, Cooper, Fallawalder, Fall Pippin, Golden Russet, Jersey Sweeting, Monmouth Pippin, Porter, Rambo, Rhode Island Greening, Wine.

"WINTER VARIETIES.—Baldwin, Black Apple, Cannon Pearmain, Danver's Winter Sweet, German Pippin, Jonathan, Lady Apple, London Sweet, Michael Henry Pippin, Newtown Spitzenburg, Roman Stem, Ortley, Pryor's Red, Rambo, Rome Beauty, Swaar, White Winter Pearmain, Wine Sap, Yellow Bellflower, Yellow Newtown Pippin, White Pippin, Black Gilliflower.

"The 'Northern Spy' and a few other celebrated varieties give fair promise of doing well here."

This apple needs a rich soil, high culture and constant growth, to produce fair fruit, as the tree grows old

Mr. A. H. Ernst, one of our most zealous and experienced pomologists, recommends, for a limited selection, 14 kinds of apples for the vicinity of Cincinnati and Southern Ohio, viz:

Early Red Margaret, Sweet Bough, Prince's Early Harvest, Summer Rose, Fall Pippin, Newtown Spitzenburg (or Ox Eye), Yellow Bellflower, Woolman's Long, White Bellflower (or Detroit of the West), Golden Russet, Broadwell Sweet, Winesap, Yellow Newtown Pippin.

Kirtland and Elliott recommend the following varieties:

SUMMER.—For the garden, Summer Rose, Early Harvest, Red Astrachan, American Summer Pearmain, Early Joe, Lowell. For market, White Juneating, Red Astrachan, Early Harvest, Williams, Red Quarrenden, Lowell.

FALL.—For the garden, Gravenstein, Fall Pippin, Fall Strawberry, Pomme Royale, Porter, Jersey Sweeting, Fameuse, Fall Harvey, Maiden's Blush, Rambo, Fall Seek-no-further, Fall Wine. For market, we prefer these to showy inferior fruits.

WINTER.—For the garden, Belmont, Swaar, Old Nonsuch, Hubbardston Nonsuch, Jonathan, Peck's Pleasant, Rhode Island Greening, Putnam's Russet (Roxbury Russet, ED.), Westfield Seek-no-further, Wine, Danver's Winter Sweeting, Wood's Greening, Tewksbury Winter Blush, Lady Apple, Fort Miami. For the market, substitute the Baldwin for Danver's Winter Sweet, and the Hollow Crown Pearmain for Wood's Greening.

Selection of apples for the vicinity of Cincinnati by a distinguished cultivator.

Red Juneating, Prince's Harvest, Summer Rose, Fall Pippin, Rambo, Newark Pippin, American Golden Russet, Newtown Spitzenburg, White Bellflower (or Detroit), Swaar, Pryor's Red, Raule's Janet, Newtown Pippin. Or these: White June, Benoni, Strawberry, Golden Sweet,

Fall Pippin, Rambo, Westfield Seek-no-further, Newtown Spitzenburg, Yellow Bellflower, Waxen (or Gate), White Pippin, Roxbury Russet. Some might wish to add the long keeper, but indifferently flavored, Gilpin or Romanite.

List of fruit for general purposes, recommended by the Cincinnati Horticultural Society.

For July and August, 2 Summer Rose; July and August, 2 Strawberry; September and October, 2 Fall Pippin; October and November, 4 Rambo; November and December, 5 Golden Russet; November and December, 5 Yellow Bellflower; November and December, 20 White Bellflower; January and February, 15 Pryor's Red; February and April, 25 Raule's Janet; March, 20 Newtown Pippin. 100 trees.

APPLES FOR INDIANA.—By a Cultivator.

Summer.—Red or Carolina June, Summer Queen, Yellow Hoss, Sweet Bough, Prince's Harvest, Kirkbridge White, Sweet June, Daniel.

Autumn.—Maiden's Blush, Wine, Holland Pippin, Rambo, Fall Harvey, Gravenstein, Ashmore, Porter.

Winter.—Black, Golden Russet, Newtown Spitzenburg, Rhode Island Greening, Hubbardston Nonsuch, Vandevere Pippin, Yellow Bellflower, White Bellflower, Michael Henry Pippin, Pryor's Red, Green Newtown Pippin, Genneting or Raule's Janet, Putnam Russet.

APPLES OF WAYNE COUNTY, INDIANA.—By a Cultivator.

Summer.—Yellow June, Sweet Bough, Sour June, Early Red, and Summer Queen.

Autumn.—Wine Apple, Fall Pippin, Rambo, and Maiden's Blush.

Winter.—Vandevere Pippin, Golden Russet, Yellow

Bellflower, Cumberland Spice, Smith's Cider, Winesap, Raule's Janet, Rhode Island Greening, Red Pearmain, Romanite, and the Butter Apple.

INDIANA APPLES.—Recommended by Henry Ward Beecher, Indianapolis.

The most popular Winter apples in Indiana, are Yellow Bellflower, White Bellflower (Detroit, of the West), Newtown Spitzenburg, Campfield, Raule's Janet (or Neverfail), Green Newtown Pippin, Michael Henry Pippin, Pryor's Red, Golden Russet, Milam, Rambo, and Vandevere Pippin, only a second or third rate table apple, but having other qualities that make it valuable to the Farmer. It seldom fails of a crop. It usually *hits* when others *miss*.

OHIO APPLES.—Recommended by S. A. Barker, McConnelsville.

Summer.—Bracken, Early Chandler, Summer Sweet (or High Topped Sweet), Pound Royal, of Marietta (Dyer), and Red Streak.

Autumn.—Rambo, Holland Pippin, Winter Russets, Yellow Bellflower, Spitzenburgs (of Marietta), Vandeveres (Red, Green, and Yellow), Red, or Long Pearmain, Red Winter Pennock, Black Gilliflower, Newtown Pippin, Westfield Seek-no-further, Rhode Island Greening, Romanite, Rome Beauty, Cooper, Orange (or Golden Sweet, of Columbus), Stone's Sweet, Sigler's Red.

APPLES OF THE SHAKERS.—Mercer Co., Ky.

Summer.—Striped June, Early Harvest, Carolina June, Summer Rose, Royal Pearmain, American Summer Pearmain, and Gravenstein.

Autumn.—Rambo, Queen, Fall Pippin, Golden Russet, Newtown Spitzenburg, and Bellflower.

Winter.—Raule's Janet, Pryor's Red, and Limber Twig.

APPLES OF JEFFERSON COUNTY, VIRGINIA.

SUMMER.—Yellow June, Vestal, Grab, Golden Sweet, Doctor Red, and Summer Pearmain.

AUTUMN. — Gravenstein, Rambo, Blenheim Orange, Bellflower, Fall Pippin, Cat Head, and Pound.

WINTER.—Newtown Pippin, Green Pippin, Lady Finger, Sheepnose, or American Golden Russet, Russet, Black Coal, Prior's Red, Limber Twig, Pennock, Abraham, Jenneting, Vandevere, and Smoke House.

ILLINOIS APPLES.

SUMMER.—Early Harvest, Sine Qua Non, Sweet Bough, Caroline, Red June, Sugar Loaf Pippin, Red Astrachan, Golden Sweet, American Summer Pearmain.

AUTUMN.—Rambo, Holland Pippin, Red Ingestrie.

WINTER.—Limber Twig, Milam, Raule's Janet, Roman Stem, Romanite, Winesap, Yellow Bellflower, Ortley, or White Bellflower, Baldwin.

VIRGINIA APPLES.

Abraham, Beverley's Red, Waugh's Crab, Raule's Janet, Limber Twig, Milam, Leather Coat, Brooke's Pippin, Ogleby, Prior's' Red, Shawn's Seedling, Belpre, Winter Cheese, Wellford, Vandevere, Hall's Red, Bonum, Winesap.

Sweet apples strongly recommended for stock, are:

Lady's Sweeting, Sweet June, Sweet Bough, Golden Sweet, Jersey Sweet, Bailey Sweet, Broadwell, Ramsdell, Danver's Winter, Talman's Sweet, Michael Henry, Campfield. For cider, the Harrison Campfield, Graniwinkle, Gilpin, and Hewe's Virginia Crab.

PEARS.

Abbott. Color, dark green, with reddish brown cheek; form, oblong obovate; size, 2; use, table; texture, sugary and buttery; quality, 2; season, August and September.

Remarks.—From Rhode Island.

Adele de St. Denis. Color, yellow, with some russet; form, obovate, irregular; size, 2 to 1; use, table; texture, melting, juicy; quality, 3; season, September.

Remarks.—Very little known here yet. Foreign.

Alpha. Color, pale yellowish green, with some reddish spots, then pale yellow blush; form, obovate, and a little oblong; size, 2; use, table; texture, buttery; quality, 2 to 3; season, September and October.

Remarks.—A seedling, from Belgium, of Van Mons. It is a moderately pleasant and passable fruit.

Amire Joannet. Color, green and yellow; form, pyriform; size, 3; use, table; texture, buttery; quality, 2 to 3; season, June and July.

Remarks.—Synonymous with Early Sugar and St. John. Exhibited before the Cincinnati Horticultural Society, by Mr. McWilliams, Nurseryman and Pomologist.

Althorpe Crassane. Color, pale green; form, roundish obovate; size, 2; use, table; texture, buttery; quality, 2; season, October and November.

Remarks.—The quality of this pear is not always

equal. But generally it may be pronounced "very good." It is of foreign origin.

AMBRETTE. Sometimes called *Tilton.* Color, green; form, oblong; size, 2 to 3; use, table; texture, juicy, buttery; quality, 2; season, November to February.

REMARKS.—Resembles, a little, Echasserie. It is a good pear.

AMBROSIA. *Echasserie by some.* Color, green; form, roundish; size, 3; use, table; texture, buttery; quality, 2; season, January.

REMARKS.—Foreign. Hardly worthy of cultivation.

Amande Double, or Amanda's Double. Color, yellow and red; form, pyriform; size, 2; use, table; texture, coarse; season, August.

REMARKS.—Foreign. Unworthy.

ANANAS. Color, clear yellow, with small dots; form, obtuse pyriform; size, 1; use, table; texture, juicy; quality, 2; season, August.

REMARKS.—Flesh, white, fine grained, firmer than the Bartlett, but of rich, sweet, and excellent flavor. Described in Downing's *Horticulturist*, "Fine flavor."

ANANAS D'ETE. Color, dull yellowish green; form, oval; size, 2; use, table; texture, sweet, juicy, pleasant; quality 2; season, August.

REMARKS.—"Very good."—*J. B. Eaton, Buffalo.*

ANDREWS. Color, yellowish and green, with a red cheek; form, roundish oval, pyriform; size, 1; use, table; texture, buttery; quality, 2 to 1; season, September.

REMARKS.—An early bearer; American origin. Good

for market. Early bearer, and productive. Exhibited by Wm. Heaver, Nurseryman and Pomologist, at the Cincinnati Horticultural Society's Rooms in 1855. Rots at the core. Bears freely.

ANGLETERRE. Color, green to yellow; form, pyriform; size, 1; use, kitchen; texture, buttery: quality, 1; season, September.

REMARKS.—Productive, though rather apt to rot soon. Beurre d' Angleterre, of Wm. Heaver; large, productive, baking.

Angora.

REMARKS.—Believed to be the Pound, or Uvedale's St. Germain. Hardy, productive, large, but good only to sell.

Aston Town. Color, greenish yellow; form, roundish; size, 3; use, table; quality, 3; season, September.

REMARKS.—From England. Generally considered unworthy.

AURATE. Color, pale yellowish green; form, regular; size, 3; use, table; texture, buttery; quality, 2; season, July.

REMARKS.—A pretty good fruit.

Autumn Colmar.

REMARKS.—Unworthy.

Autumn Bergamot. Color, brownish green, yellow when ripe; form, flat at blossom end, stem short; size, 2; use, table; texture, juicy; quality, 3; season, September.

REMARKS.—Tree, not vigorous, but prolific. Unworthy.

AUTUMN PARADISE. Color, dull yellow; form, obovate, acute pyriform; size, 1; use, table; texture, buttery, melting, juicy; season, September and October.

REMARKS.—Of foreign origin. Similar, in most respects, to Beurre Bosc.

AUTUMN SUPERB.

REMARKS.—Large, obtuse, pyriform. Good for kitchen use. A great bearer.

Barker. Color, greenish yellow; form, obovate; size, 2; use, table and baking; texture, coarse; quality, 2; season, September and October.

BARON DE MELLO. Color, yellow and russet; form, obovate, acute pyriform; size, 2; use, table; texture, melting, juicy; quality, 1; season, September and October.

REMARKS.—"Good and new."—*Horticultural Exhibition*, 1855. "Very good."—Marshall P. Wilder, *in Horticulturist.*

BARTLETT, or *Williams' Bon Chretien*, with a great number of synonymes (showing its great value everywhere). Color, yellow; form, obovate, obtuse pyriform; size, 1; use, table; texture, buttery, melting, sugary, juicy, with a peculiar flavor; quality, 1; season, August to October.

REMARKS.—English. Originated in 1770. Tree vigorous, and very early productive. "Well suited to the vicinity of Cincinnati."—*F. G. Cary.* Excellent, early, productive, either on pear or quince. The former stock sufficiently good. Thus far, taken altogether, the best pear of its season for the locality of Cincinnati. It is even now too little grown, but continually gaining the great favor and estimation it so justly and fully merits. It can be grafted, and do well even on the apple. The Bartlett is a pear that it is not necessary to cultivate on the quince. On its own stock it does not grow large to take up much room, and from its very productive character naturally, it bears too full on the quince. It is one of those rare pears that succeeds equally well North and South. It takes the widest range of climate possible for the pear kind. This is the pear for this vicinity. It need not be grafted on the quinces, for it bears young enough on its own roots. It is a thrifty grower, produces the second year from the graft, when put on large trees. We have had it to bear the first year. It outsells any thing else. Three dollars per bushel is the usual price—often more.

Beadnell. Color, pale yellow and green; form, turbinate; size, 2; use, table; texture, melting, and very juicy; season, September.

REMARKS.—Foreign.

BELLE EXCELLENT.—Color, yellow, with a red blush; form, oblong pyriform; size, 1; use, table; texture, melting, buttery; quality, 1; season, September.

BELLE DE BRUSSELLES, or *Beauty of Brussels.* Color, deep yellow; form, variable, obovate pyriform; size 1; use, table; texture, sugary; season, August.

REMARKS.—Exhibited by W. S. Hatch, August, 1855. Fruit Committee considered it a first-rate market fruit, and of pretty good flavor. This is presented as Belle of Flanders (erroneously) in the Catalogue of the London Horticultural Society.

Belle Canaise. Color, light yellow; form, obovate; size, 3; use, table; texture, juicy, coarse; season, October to January.

REMARKS.—Foreign. Unworthy.

BELLE OF FLANDERS, or ***Flemish Beauty, Bosch,***

Bouche Nouvelle, Bosc Sire, etc., etc. Color, pale yellow, mostly covered with marblings and patches of light russet, and reddish brown in the sun; form, oblong, obtuse obovate; size, 1; use, table; texture, not very fine grained, juicy, melting, sugary; quality, 1; season, Aug. and Sept.

REMARKS.—Very much admired, and the flavor greatly approved of. Considered to rank among the very best. "Merits the first place with the Bartlett, Seckel, etc., among the most delicious pears tested in this neighborhood." "One of the best."—*Dr. Warder.* Very fine specimens exhibited by the author, August 25, 1855. It is deserving of the most general cultivation. It succeeds well on the quince. The rich soils of the West suit it. Tree vigorous, with the branches upright, and shoots dark brown. Has been sometimes mistaken for Knight's Monarch. This comes next to the Bartlett, but does not bear so young. Delicious. Rarely handsome.

BELLE LUCRATIVE. See Fondante d'Automne. A most delicious fruit. A universal favorite.

BELMONT.

REMARKS.—Large. Good for cooking. A good bearer, and profitable.

BELLE JULIE. Color, yellowish green; form, long ovate; size, 2; use, table; texture, juicy; quality, 2; season, September and October.

REMARKS.—"Very good."—*Hovey's Magazine.*

BELLE DE NOEL, *or Belle Apres Noel.* Color, bright yellow; form, obovate, obtuse pyriform; use, table; texture, juicy; quality, 2; season, December.

REMARKS.—Very rich, and highly flavored. A great keeper. "Good keeper."—*Dr. Warder.*

BELLISSIME D'ETE, or the *Beauty of Summer, or French Jargonelle, Red Muscadel, English Red Cheek, etc., etc.* Color, yellow, with red cheek; form, roundish obovate; size, 3; use, table; texture, juicy, sugary; season, July, and sometimes August.

REMARKS.—The fruit is small and singularly beautiful; the skin is smooth, of a bright yellow, the cheek toward the sun of a brilliant red, with small dots; the form is regular, diminishing toward the stem, which is long. If picked before it is ripe, it is a pretty good early pear; it sometimes grows in clusters; produces abundantly, and commonly ripens about the middle of July

BELLE DE THOUARS. Color, brownish russet; form, pyriform, angular; size, 2; use, table; texture, tart, juicy; quality, 1 to 2; season, August and September.

REMARKS.—Of foreign origin.

BELLE ET BONNE, or *Beautiful and Good, or Gracieuse.* Color, greenish yellow; form, roundish; size, 2; use, table; texture, juicy and melting; quality, 2; season, August.

REMARKS.—Pretty good.

BENSELL'S WINTER. Color, yellow; form, round; size, 1; use, baking; texture, juicy, astringent; quality, 2; season, Winter.

REMARKS.—Originated near Philadelphia; a fine keeper and a great bearer. Somewhat harsh and astringent, though not of a bad quality for cooking.

BENOIST NOUVEAU. Color, greenish yellow; form, round obovate; size, 2; use, table; texture, juicy, sugary; quality, 2; season, Winter.

REMARKS.—Also rather astringent.

Bergamotte d'Esperen.

Remarks.—A great keeper. Like the Autumn Bergamotte, grows well on quince. Melting and juicy. Not quite so rich as some of the Autumn Pears.

Bequesne. Color, yellow, with dark spots; form, long; size, 1; use, baking; texture, astringent; quality, 2.

Remarks.—Only good for cooking.

Bergamotte Sylvanche. Color, green, size, 2; use, table; texture, juicy; quality, 2; season, August and September.

Beurre de Capiaumont. Color, clear yellow; form, long turbinate; size, 2 to 3; use, table; texture, crisp; quality, 2 to 3; season, August and September; situation or aspect, South.

Remarks.—A great bearer. Exhibited by the author at the Cincinnati Horticultural Society Hall, August 25, 1855. Committee considered it "a pretty good pear, but subject, when the trees bear full, to be a little too crisp and astringent." These specimens, more astringent than it usually is, and, therefore, not so good as the pear sometimes is. An annual bearer. A good little pear; when ripened in the house, it loses its astringency. Too small for market.

Beurre de Beaumont. Color, yellowish green; form, round obovate; size, 2.

Remarks.—Foreign. Good.

Beurre Preble. Color, greenish yellow; form, oblate obovate; size, 1; use, table; texture, buttery; quality, 2; season, September.

Remarks.—Not very good.

BEURRE MOLLETTS GUERNSEY. Color, yellowish green; form, ovate pyriform; size, 2; use, table; texture, melting; quality, 2; season, November to December.

REMARKS.—Foreign. A fine kind.

BEURRE BENOIST. Color, yellow, mottled; form, obtuse pyriform; size, 2 to 1; use, table; texture, melting, buttery, sweet and rich; quality, 1; season, July.

REMARKS.—A new and exceedingly fine pear; always fine. Flesh, white, tender, buttery, abounding in rich, sweet, sprightly juice.

BEURRE D'ANJOU. Color, pale yellow, with a dull blush; form, oblong, obovate pyriform; size, 1; use, table; texture, juicy, melting; quality, 1; season, October and November.

REMARKS.—Foreign. Good on pear and quince. An old pear, on Loudon's List. "Fine, good, rather acid."—*Cincinnati Horticultural Exhibition*, 1855. Of nearly the highest excellence. A great many good qualities about it.

BEURRE D'AREMBERG. Has many synonymes. Form, obovate, obtuse pyriform; size, 2; use, table; texture, juicy, melting, vinous; quality, 2; season, December to February.

REMARKS.—Foreign. Sometimes confounded with Glout Morceau, and Soldat Laboreur. Fruit hangs well. Vigorous and hardy trees. A warm, rich soil suits it. Wood, strong, long jointed. Rather difficult to ripen: a common fault with Winter fruit. A warm temperature has been found injurious for this purpose. A cool place, about 40° or 50°, and taking a long time gradually to ripen, appears, by late experiments, to be the most successful method for this purpose. The Buerre d' Aremberg can not be called a good Winter pear with us,

West. It has the fault of being too long in coming into bearing.

Beurre Easter. See Easter Beurre.

Beurre Rhine. Color, light yellow, rough spots; form, pyriform, irregular; size, 1; use, table; texture, rather coarse; quality, 2; season, October and November.

Remarks.—Succeeds very well on the quince. Often very good in the East. "Poor," by the Fruit Committee Cincinnati Horticultural Exhibition, 1855.

Beurre Brown, or Brown Beurre. Color, yellowish green on brownish ground; size, 2; use, table; texture, buttery, melting, juicy; quality, 2; season, September.

Remarks.—An old variety. Requires a warm, rich soil. A peculiar vinous taste. Much finer in England than in the West. There it is one of their choicest fruits.

BEURRE BOSC. Color, dark yellow, with russet dots; form, obovate, acute, pyriform; size, 1; use, table; texture, juicy, melting, sweet; quality, 1; season, September and October.

Remarks.—Fruit, always fine. Foreign, by Van Mons. Fruit varies somewhat in size. Tree, vigorous; long, brownish olive shoots. Very fine, though a wild grower. Exhibited by T. M. Millikin, of Hamilton county, at the Cincinnati Horticultural Exhibition, 1855. Will not do on the quince stock; perhaps the only one that will not do at all. Varies in different seasons. Very acid and rough, sometimes. We do not think it suits our climate.

BEURRE SUPERFIN. Color, dull pale green; size, 1; use, table; texture, juicy; quality, 2; season, September and October.

REMARKS.—Excellent in Boston; received prize. Flesh, juicy, melting, and with slight aroma. Described by Col. Wilder, in *Horticulturist.*

BEURRE SPRIN. Form, obovate obtuse pyriform; size, 1; use, table; texture, juicy; quality, 2; season, October.

REMARKS.—Flesh, melting, juicy, rich with a peculiar aroma. Described by Col. Wilder.

Beurre Bachelier. Color, greenish yellow; form, oblate obovate pyriform; size, 1; use, table; texture, sugary; season, Winter.

REMARKS.—Foreign.

BEURRE D'AMALIS. Color, dull green; form, obtuse pyriform; size, 1; texture, juicy; quality, 2; season, August.

REMARKS.—Rots at core before it looks ripe. Fine in flavor, however, when sound. Exhibited by A. H. Ernst, August, 1855. Committee considered these specimens "very fine; a first-rate fruit." Not much cultivated, yet, in this vicinity. If it generally rots, which is likely, it will be better to regraft old trees, and cease cultivating it. Committee must have been in error about this fruit; not a first-rate fruit. Good on alternate years, when it bears heavy crops. It is too small for market, because it comes in with the great Bartlett. Specimens of the same fruit will often vary very much—so much so as often to deceive the most experienced. It sometimes requires several trials of the same fruit to be infallible.

BEURRE DE WATERLOO. Color, dull green, rough skin, with russet traces and points; form, obtuse pyriform; size, 1; use, table; texture, juicy, sprightly, sugary; quality, 1; season, September.

Remarks.—Described by Col. Wilder in *Horticulturist.* Fine with us (Cincinnati). Exhibited in Mr. Ernst's collection, at the Horticultural Exhibition of 1855.

Beurre Steikman. Color, dull grayish russet; form, obovate pyriform; size, 2; texture, rich, sub-acid; season, October.

Buerre Brettonneau. Form, obovate oblate pyriform; size, 1; use, table; texture, sugary, melting; season, long keeper.

Remarks.—Large and handsome. Flesh, melting, high flavored and excellent. Described thus briefly by Col. Wilder, President of Massachusetts Horticultural Society for many years. Foreign. "Does not succeed well on the quince."—Rivers, in *Horticulturist.*

Buerre Diel. Color, yellow; form, obovate, obtuse pyriform; size, 1; use, table; texture, juicy; quality, 2; season, October to November.

Remarks.—Fine when well ripened. Thrifty, fruit roughish. Loses its leaves badly. Foreign. Does well on pear or quince, but best on quince. Very productive. Difficult to ripen, and leaves apt to fall, although an Autumn fruit. Fine, large, very delicious when ripened in the house, as most pears should be. Very subject to leaf blight. We kept them until January one season, when they were pronounced very superior by the members of the Cincinnati Horticultural Society. Not as great a bearer as the Bartlett.

Beurre Clairgeau of Nantes. Color, yellowish green; form, irregular turbinate; size, 1; use, table; texture, buttery; quality, 1; season, October and November.

Remarks.—Described by Andre Leroy in *Horticulturist.*

A handsome pear, of first-rate quality. Flesh, melting, juicy. Resembles Gray Doyenne. Very productive.

Beurre Charron. Color, greenish yellow; form, roundish; size, 2; use, table; texture, juicy, melting; quality, 1.
REMARKS.—Foreign.

Beurre Kenrick. Color, greenish yellow; form, pyriform; size, 2; use, table; texture, not juicy.
REMARKS.—Unworthy.

BEURRE NANTAIS. Color, greenish yellow with crimson; form, long pyriform, sometimes obovate; size, 2; use, table; texture, melting, very juicy; richly flavored, sweet and pleasant; quality, 1; season, August.

REMARKS.—Should be picked before ripe. Summer and early Fall pears should be picked when fully grown, and before the process of ripening commences. Much fine fruit is spoiled by being picked too late. With very few exceptions, no pear ought to ripen on the tree.

Beurre Knox.
REMARKS.—Unworthy.

BEURRE PREBLE. Color, greenish yellow, russet and green spots; form, oblong, obovate; size, 1; use, table; texture, buttery; quality, 2; season, September and October.
REMARKS.—American origin. Worthy.

Beurre Colmar.
REMARKS.—Unworthy.

BEURRE LANGLIER. Color, light green, with pale yellow; form, obovate, pyriform; size, 1; use, table;

texture, juicy; quality, 1; season, December and January.

REMARKS.—Flesh, yellowish white, melting, and fine-grained. Flavor, sprightly, sub-acid, rich, excellent, with a light perfume. Described by Col. Wilder, in *Horticulturist.*

BEURRE DE BEAUMONT. Color, yellowish green, brownish red in sun, with many dark green or russety spots; form, roundish obovate; size, 2; use, table; quality, 2; season, September.

REMARKS.—Of foreign origin. Flesh, white, buttery, juicy, sweet, and pronounced by good authority, "very good."

BEURRE GOUBAULT. Size, 1; use, table; quality, 1; texture, buttery; season, August and September.

REMARKS.—Good and large. Exhibited by Wm. Heaver, at Cincinnati Horticultural Society, August 25, 1855. Committee on Fruit considered it "rich, buttery, and of very fine flavor."

BEURRE GIFFORD. Color, yellowish green; form, pyriform; size, 2; use, table; texture, juicy, melting; season, August.

REMARKS.—Foreign. A beautiful, early pear. Does well on quince stocks. Of fine texture and flavor.

Beurre Van Mons. Color, yellowish, with russet; form, pyriform; size, 2; use, table; texture, coarse, insipid to some extent; quality, 3; season, September.

REMARKS.—Exhibited before the Cincinnati Horticultural Society, August 18th, 1855. Fruit Committee pronounced it third-rate. The judgment of fruit often depends upon eating it at the exact point of perfection.

BEURRE ANDUSSON. Color, yellowish green when ripe; form, obovate, acute pyriform, tapering abruptly; size, 2; use, table; texture, juicy, tender; quality, 2; season, August and September.

REMARKS.—Flesh, melting, tender, juicy. Flavor, rich sub-acid, slightly perfumed with rose. Described by Col. Wilder, in *Horticulturist*.

Beurre Moir. Color, pale green; form, obovate, oblate, pyriform; size, 1; use, table; texture, juicy, vinous; season, September and October.

REMARKS.—Foreign. Not tested here.

Beurre Romain. Color, yellowish green; form, obovate; size, 2; use, table; texture, poor; quality, 4; season, September.

REMARKS.—Unworthy. Very poor.

Beurre Gris. See Brown Beurre.

BEURRE THOURY. Color, greenish yellow; form, round, obtusely turbinated; size, 2 to 1; use, table; texture, buttery, juicy; quality, 2; season, August and September.

REMARKS.—Tree, handsome and vigorous. "Almost, very good." Exhibited before Cincinnati Horticultural Society, 1855, by Elliott, of Cleveland, author of *American Fruit Grower's Guide*, a valuable work, but too general in its character.

BEURRE DE RANZ, *Beurre Epine*, *Beurre de Flandres.* Color, dark green, with a sunny bronze, some russet at the crown, with some russet dots; form, oblong, obtuse pyriform; size, 1; use, table; texture, coarse, juicy; quality, 2; season, January to March.

REMARKS.—Varies a great deal in quality. Not often "very good." Tree, very straggling in habit. Foreign.

BEURRE BEAULIEU. Color, yellow, with russet spots; form, obovate pyriform; size, 2 to 1; use, table; texture, juicy; quality, 2; season, August and September.

BEURRE CHARRON. Form, round, obovate; size, 2; use, table; texture, juicy; quality, 1; season, August and September.

REMARKS.—Flesh, exceedingly melting, juicy, and perfumed. A new pear, from Anjou, France.

BEURRE SUISSE. Color, green, red and yellow ground; form, obovate oblate; size, 2; use, table; texture, poor, tasteless; quality, 3.

REMARKS.—Foreign. Only curious.

Beurre Bronzee.

REMARKS.—Unworthy.

BEURRE OSWEGO. Color, yellowish green, dull, blotches of russet; form, ovate obovate, or obovate rounded; size, 2; use, table; texture, melting, juicy, sub-acid, sprightly; quality, 2; season, September.

REMARKS.—This is of value, but size is small. Very juicy. Of brownish color. Tree said to be hardy, September, or sometimes October. We do not find this equal to its reputation.

BEURRE FIGUE. Color, greenish yellow; form, round; size, 2; use, table; texture, tender, juicy; quality, 1; season, August.

REMARKS.—A small pear. "Good."—*A. H. Ernst*, nurseryman, who has paid particular attention to the pear.

BEURRE VAN MARUM. Color, greenish yellow; form, obtuse pyriform; size, 1; use, table; texture, tender; season, October.

REMARKS. — Productive and fair. An early bearer. Good for market.

BEURRE SPENCE. See Flemish Beauty.

BEURRE CRAPAUD. Color, deep yellow; form, round obovate; size, 2; use, table; texture, melting, juicy; quality, for Cincinnati 3.

REMARKS.—Fine for the North. Foreign. Vigorous upon both pear and quince.

BEURRE NANTAIS. Color, pale yellow, with some dots of russet; form, oblong pyriform; size, 1; use, table; texture, melting, juicy; quality, 2; season, August and September.

REMARKS.—Described in *Hovey's Magazine.*

BEURRE SUPERFIN. Color, pale green, traces of russet, cheek of a dusky brown; size, 1; use, table; texture, juicy, melting, slight aroma; quality, 2; season, Sept.

REMARKS.—Col. Wilder in the *Horticulturist.*

BEURRE PICQUERY. Color, pale yellow, with gray dots; form, obovate pyriform; size, 2; use, table; texture, melting, buttery, and juicy; quality, 1; season, September to November.

REMARKS.—Very rich and good. Of great excellence. Horticultural Exhibition, 1855. Urbaniste, according to Elliott. Suited to rich soils, West.

BEURRE GRIS D'HIVER NOUVEAU. See Glout Morceau. Winter, one of the very best.

BEURRE KOSSUTH. Color, dull yellowish green; form, turbinate rounded; size, 1; use, table; texture, juicy, sugary; quality, 2.

REMARKS.—Foreign. New. A little acid. Described in *Hovey's Magazine*.

BEURRE MILLET. Color, pale yellow; form, rounded, obtuse pyriform; size, 2; use, table; texture, melting, juicy; quality, 1; season, September.

REMARKS.—Exhibited by A. H. Ernst in the Autumn Fair of the Cincinnati Horticultural Society, 1855. Very fine.

Beurre Vaet. Color, greenish yellow, brown cheek; form, obovate; size, 2; use, table; texture, indifferent; quality, 3; season, October.

REMARKS.—Unworthy of attention here.

Beurre d'Heri. Color, greenish yellow; form, roundish; size, 2; use, table; texture, poor; quality, 3; season, September.

REMARKS.—Not worthy of any attention.

Bergamotte Suisse. Unworthy.

Bergamotte d'Esperen. Color, yellowish green; form, roundish; size, 2; use, table; texture, buttery, melting, juicy; quality, 1; season, Winter.

REMARKS.—Foreign.

BERGAMOTTE GAUDRY. Color, yellowish green; form, roundish; size, 2; use, table; texture, tender, juicy; quality, 2; season, Winter.

BERGAMOTTE CADETTE. Color, pale green; form, round

obovate; size, 2 to 3; use, table; texture, tender, juicy, a little gritty; season, September.

REMARKS.—Foreign, as, indeed, all these French names imply.

BEZI DE CHARMONTELLE, or *Winter Butter*. Color, yellow, reddish next the sun; form, very irregular; size, 1; use, table; texture, melting; quality, 2; season, long keeper.

REMARKS.—Crown of the fruit very deeply hollowed.

Bezi de Caissoy. Color, greenish yellow, when ripe; form, round, flat at crown; size, 2 to 3; use, table; texture, buttery; quality, 2; season, October.

BEZI DE MONTIGNY. Color, yellowish green; form, obovate; size, 2; texture, juicy, sugary, tender; season, September.

REMARKS.—Nice little pear. Of not much flavor, mushy.

BEZI SANSPAREIL. Color, dull greenish yellow; form, obtuse, pyriform; size, 3; use, table; texture, juicy, aromatic; quality, 2; season, Winter.

REMARKS.—Of foreign origin.

Bezi de la Motte. Color, dull green; form, round; size, 1; use, table; texture, buttery; quality, 2; season, October.

REMARKS.—An old variety, which appears as hardy as deficient in high flavor; sweet and juicy. Productive, early.

BEZI DE NAPLES. Color, light yellowish green; form, ovate, obovate; size, 2; use, table; texture, buttery, juicy, sugary; season, September.

Bishop's Thumb, or Beurre Adam. Unworthy.

BLOODGOOD. Color, yellow; form, obtuse, pyriform; size, 2 to 3; use, table; texture, buttery; quality, 1; season, July, August.

REMARKS.—Very fine, though rather small. The best early pear, but not the earliest. Exhibited by several during the season. The Fruit Committee decided it a first-rate Summer fruit.

Bleeker's Meadow, or Pfiester. Color, yellowish; form, roundish; size, 2; use, table; texture, medium. "Pretty good."—*J. A. Warder.* Mr. Ernst thinks it, or has found it, rather unworthy. Juicy, but not rich.

Black Worcester. Color, russet green; form, obovate; size, 1; use, kitchen; texture, harsh; quality, 2; season, November to February.

REMARKS.—Flesh, firm, coarse, austere. For baking only.

BON CHRETIEN FONDANTE. Color, yellowish green; form, round, oblate; size, 2; use, table; texture, melting, juicy, tender, sugary; quality, 2.

REMARKS.—A little gritty, but pretty fine. A very melting, good pear; not very rich, but a uniform bearer; nearly always smooth.

BON CHRETIEN D'HIVER, or *Good Christian of Winter.* Color, yellow; form, truncated pyramidal; size, 1; use, table; texture, tender, sugary; season, January.

REMARKS.—Flesh a little odoriferous. Size, sometimes six inches long and four wide.

BONNE DES LEES. Color, light yellow, and red; form,

obovate pyriform; size, 2; use, table; texture, melting, juicy; quality, 1; season, September.

BORDENIEVE. Color, dull greenish russet; form, acute pyriform; size, 2; use, table; texture, juicy, buttery, sugary; season, September.

Boucguia. Unworthy.

BON CHRETIEN D'ESPAGNE. Color, yellow, brown dots; form, long; size, 1; use, table; texture, juicy; quality, 2; season, November.
REMARKS.—Requires a good soil.

Brocas Bergamot. Color, dull light green; form, round; size, 2 to 1; use, table; texture, juicy; season, September.
REMARKS.—An uncertain and small bearer; deficient in vigor. Loses its leaves early.

BOUSSOUCK. Color, yellow, russet spots; form, globular, obtuse pyriform; size, 1; use, table; texture, juicy; season, October.
REMARKS.—Introduced by Kenrick. Tree vigorous.

BROWN BEURRE, or *Beurre Gris of Coxe* (the best and most practical writer of his time on fruit in the Middle States). Color, green, with black clouds; form, almost elliptical; size, 1; use, table; texture, juicy; quality, 2 in this country, 1 in England; season, September.
REMARKS.—Requires a very rich soil; varies in excellence; sometimes too acid, sometimes cracks when the year is unfavorable. It lasts a long time in favorable seasons. This was the great pear of our boyhood in England, where it is superlatively fine, rich, and melting, like the Bartlett.

Broome Park. Color, brown; form, roundish; size, 2; use, table; texture, rather poor; quality, 3; season, November.
REMARKS.—Unworthy.

BRANDYWINE. Color, dull yellow and green; form, tapers to stalk; size, 2; use, table; texture, juicy; quality, 1; season, August, same time as the famous Bartlett.
REMARKS.—Flesh more rich, juicy, and sprightly than the Bartlett; not so brilliant in complexion. It never rots in core, and keeps well rather early picked from the tree. Tree very thrifty. Of American origin.

BRANDE'S ST. GERMAIN. Color, bright yellow; form, oblate oval; size, 2; use, table; texture, juicy, vinous; quality, 2; season, Winter.
REMARKS.—Foreign. Manning thinks well of it, considering its season.

BRINGEWOOD. Color, yellow, russet and brownish; form, pyriform; size, 2; use, table; texture, rich; quality, 2; season, November.
REMARKS.—Of English origin. Gritty at core, though otherwise well flavored.

Brielmont. Color, yellow; form, obovate oblate; size, 3; use, table; texture, melting; quality, 2; season, September and October.
REMARKS.—Of foreign origin.

BROUGHAM. Color, yellowish white; form, obovate; size, 3; use, table; texture, buttery, sugary; quality, 2 to 1; season, Fall.
REMARKS.—Foreign. A little gritty. This is objectionable, but many very rich pears, as Gansell's, Bergamot, and a few others, have this defect.

BUFFUM. Color, brown; form, ovate; size, 2; use, table; texture, buttery; quality, 2; season, August and September.

REMARKS.—Of American origin. Productive and early. Native of Rhode Island. Excellent for standard orcharding. An upright, strong grower. Always productive. A great fruit for marketing.

Burnett. Unworthy.

CABOT. Color, russet yellow, red in sun; form, oval, roundish; size, 2; use, table; texture, hardly passable; quality, 3; season, August.

REMARKS.—"A good little fruit."—*A. H. Ernst.*

Caen de France. Color, yellow russet; form, round obovate; size, 2; use, table; texture, juicy, sugary; season, Winter.

REMARKS.—Foreign.

CAPSHEAF. Color, yellow, with a great deal of common russet; form, roundish obovate; size, 2; use, table; texture, melting, juicy, sweet; quality, 2; season, September.

REMARKS.—Native of Rhode Island. Tree, hardy.

Calabasse. Unworthy.

Calabasse Grosse. Unworthy.

Capucin. Unworthy.

CAPIAUMONT. Color, yellow; form, globular, acute pyriform; size, 2; use, table; texture, buttery, sugary; quality, 2; season, September and October.

REMARKS.—Good on pear and quince. A free grower.

CALHOUN. Color, yellow; form, roundish; size, 2; use, table; texture, melting, juicy, sugary; quality, 2; season, September and October.

REMARKS.—American. New Haven, Connecticut.

CANANDAIGUA. Size, 1; texture, juicy, sugary; quality, 2; season, August.

REMARKS. — Resembles the Bartlett. Does well on quince or pear. Tree, healthy, vigorous and productive. Described in *Hovey's Magazine.*

Calebasse d'Ete. Color, dull green; form, oblate, pyriform, irregular; size, 2 to 1; use, table; texture, buttery, juicy; quality, 2; season, August.

CATILLAC. Color, yellow; form, obtuse pyriform; size, 1; use, kitchen; texture, buttery, juicy; quality, 2; season, November to March.

REMARKS.—For Winter baking.

CATINKA. Color, yellow; form, oblate pyriform; size, 1; use, table; texture, juicy, sugary; quality, 1; season, September and October.

REMARKS.—New and good. Large, round, coarse baking pear. Good bearer.

CHANCELLOR, or *Early German.* Color, green; form, oblate obovate pyriform; size, 1; use, table; texture, melting; quality, 1; season, August and September.

REMARKS.—Brinckle (one of the best authorities), in *Horticulturist*, pronounces this a truly delicious pear.

CHAPTAL. Color, greenish yellow; form, round obovate ovate; size, 2; use, table; texture, juicy; quality, 1; season, Winter.

Charles Van Mons. Color, yellowish green, russet spots; form, obovate obtuse pyriform; size, 3; use, table; texture, juicy, melting, vinous; quality, 2; season, January.

Remarks.—Of foreign origin, as the name denotes; Van Mons, the great originator of some of the finest pears from seed, having used the greatest perseverance and energy, skill and science, in his labors.

Chelmsford.

Remarks.—Like the Stone, is good to sell and to bake, not to eat for the dessert. Its fine appearance promises, like many things, more than it can perform, and attracts the popular eye.

Charles d'Austria. Unworthy.

Chaumontel Tres Gros, or *Chaumontel.* Very rich and large. See Beurre Easter, or Easter Beurre, it being synonymous.

Chaumontel, or *Winter Beurre.* Color, yellowish russet brown in sun; form, oblate obovate; size, 1; use, table; texture, buttery, melting; quality, 1; season, October to February.

Remarks.—Requires a rich, warm soil.

Christmas. Color, bronzed, russety; form, ovate, roundish; size, 2; use, table; a little gritty, but juicy and sugary; quality, 2 to 1; season, November to February.

Remarks.—American. New. From this vicinity (Cincinnati). *Warder's Notes.*

Citron. Color, dull green, with russet spots; form, roundish obovate; size, 2; use, table; texture, somewhat coarse, but melting and juicy; quality, 2; season, August.

CITRON DES CARMES, or *Madeleine.* See Madeleine. The best and most profitable, and earliest pear of any great merit. Good, if picked before it turns yellow. Early, medium; not at all like Green Chisel, as has been said.

Clara. Unworthy. Inferior.

CLION. See *Vicar of Winkfield.*

Clinton. Unworthy.

COLLINS. Color, yellowish green; form, round, obovate; size, 1; use, table; texture, vinous, juicy, sugary; quality, 1; season, August.

REMARKS.—From Watertown, Massachusetts. Supposed to be a seedling of White Doyenne (that most excellent fruit). A great bearer.

COLMART. Color, a little yellowish when ripe, green and brown spots; form, flat at blossom end; size, 1; use, table; texture, buttery, melting; quality, 2; season, Winter.

REMARKS.—Keeps very well.

COMPTE DE LAMY. Color, yellow, with a brown cheek, with russet dots; form, round, obovate; size, 2; use, table; texture, buttery; quality, 2; season, August and September.

REMARKS.—Exhibited by Heaver & Eyler, August 25, 1855. Rich, delicious, and sugary. Tree of upright growth. Well worthy of cultivation.

Comprette. Unworthy.

Commodore. Inferior.

CONSEILLER RAMUEZ. Color, dull greenish russet; form, obovate, obtuse pyriform; size, 2; use, table; texture, melting, tender; quality, 2; season, September.

REMARKS.—Described and recommended by Col. Wilder, in *Horticulturist.*

Colmar Epine. Color, dull green; form, round, obtuse, oblate; size, 1; use, table; texture, melting, vinous, juicy; quality, 1; season, August and September.

REMARKS.—Origin foreign.

Colmar Neill. Color, pale yellow; form, obovate; size, 1; use, table; texture, buttery, melting; quality, 2; season, September.

REMARKS.—Origin foreign.

COLUMBIA. Form, oblate, obovate pyriform; size, 2; use, table; texture, juicy; quality, 2; season, Winter.

REMARKS.—From Westchester, New York. Tree very hardy and productive. Fruit always smooth and fair, but not very good. It is, however, a good keeper.

Comstock. Inferior.

COLMAR D'AREMBERG. Color, green on deep yellow; form, irregular; size, 1; use, table; texture, crisp; quality, 2 to 3; season, October and November.

REMARKS.—Skin, thin. Flesh, neither fine nor coarse. Juice, abundant, acerb, but good, nevertheless, or rather, middling. Only "so-so" in flavor. It has sometimes been confounded with the Glout Morceau. Productive.

Colmar. An indifferent pear.

COTER. Color, pale green; form, regular obovate; size, 2; use, table; texture, tender, melting; quality, 2; season, October and November.

REMARKS.—Tree, healthy and vigorous.

COUNTESS OF LUNAY. Color, pale yellow, red and russet; form, roundish obovate; size, 2; use, table; texture, crisp, melting, juicy; quality, 1; season, September.

REMARKS.—Tree, vigorous; very productive on quince. Has been distributed as the Doyenne d'Ete, in some parts of the country.

COIT'S BEURRE. Color, rich brown russet; form, obtuse pyriform; size, 1 to 2; use, table; texture, juicy, buttery, melting; quality, 1; season, August and September.

REMARKS.—From Euclid, Ohio. Tree, hardy, vigorous, upright. Well worthy a place in all collections. Ranks among the first. Can hardly be too highly commended. "Very rich, and luscious."—*Fruit Committee at the Cincinnati Horticultural Exhibition*, 1855, the great Western Fruit Year, when there were so many fruits tested, and decided upon, *pro* and *con*—a valuable and memorable year for pomologists, and the public at large.

Crawford. Unworthy.

Croft Castle. Inferior. Worthless.

CROSS, or *Winter Cross*. Color, deep yellow; form, round obovate; size, 2; use, table; texture, melting, juicy; quality, 1; season, October to February.

REMARKS.—Of American origin; from Massachusetts. Tree, hardy, slender. Wood, grayish yellow.

CRASSANE, or *Bergamotte Crassane.* Color, greenish yellow; form, rather round, with black dots; use, table; texture, melting, juicy; quality, 2; season, September.

REMARKS.—Ripens soon after the Yellow, or Orange Beurre. Dry when first gathered. It will keep in the house six weeks. It is a great bearer; of vigorous growth, and hardy.

Cuisse Madame. Color, yellowish green; form, long; size, 2; use, table; texture, juicy; season, July.

REMARKS.—Possesses a slight musky flavor. Liable to be blown from the tree.

CUSHING. Color, light, greenish yellow, with grayish dots, or patches; form, ovate obovate; use, table; texture, white, melting, sugary; quality, 1; season, August.

REMARKS.—American. From Hingham, Massachusetts. Tree, hardy. Of fine quality, as many from this State (Massachusetts), particularly near Boston, with that distinguished Pomologist and Merchant Prince, M. P. Wilder, to take the lead, with his excellent judgment, and great zeal, fully and truly demonstrate. The Cushing is good for market.

Cumberland.

REMARKS.—An indifferent fruit. Sometimes pretty good, but we would not recommend it.

DALLAS. Color, dull yellow and red, with russet; form, round, obovate; size, 2; use, table; texture, white, melting, juicy; season, October.

Remarks.—American. From New Haven, Connecticut. It may be observed, as a general rule, that the *native* fruit trees, originating, either from native seed, or from the seed of established fine foreign kinds, are the hardiest, most thrifty, vigorous, and healthy—as the Seckel, Buffum, Dearborn's Seedling, etc. The Bartlett, and some others, are exceptions to this rule.

D'Amour.

Remarks.—Rather too small to be popular here, though sometimes of rich flavor. It is a very fruitful tree, but too small to be much cultivated.

Dean's Summer. See Doyenne d' Ete.

DEARBORN'S SEEDLING. Color, pale yellow, with russet spots; form, obovate; size, 3; use, dessert; texture, juicy, melting; quality, 1; season, August.

Remarks.—Tree bears well at 7 or 8 years old. First-rate. Origin Roxbury, Massachusetts. Tree very vigorous, erect, yet spreading. Very fine and delicious, but rather small. Exhibited by A. H. Ernst, with whom it has always been a great favorite, August 11, 1855. Fruit Committee decided it, "Hardly to be too highly prized." "Very fine."—*Dr. Warder.* An early and abundant bearer. A hardy tree.

De Louvain. Color, dull greenish yellow; form, obovate, acute pyriform; size, 1; use, baking; texture, crisp, juicy, astringent; quality, 1; season, December to February, and even March.

Remarks.—Foreign. It is well known that, for the purposes of good baking or cooking, there are qualities required different from the table or dessert. These qualities are juiciness, crispness, some acidity; and even astringency, or harshness, and hardness, is but little objectionable for this purpose in most fruits. Still the best fruits, when they do cook well, retain their delicious and superior points as well after the culinary processes are applied to them, as before.

DELICES D'HARDENPONT. Color, pale yellow, green dots, russet in sun; form, roundish; size, 2; use, table; texture, white, buttery, melting, juicy; quality, 1; season, September.

REMARKS.—Foreign. We are greatly indebted to Mr. Knight, the great English botanist, for many new and most excellent fruits, obtained by cross impregnation; as well as to Van Mons and others, for kinds from the seed. It is well known that upon this first, most delicate, and beautiful process, must greatly depend our improvement, especially in grapes, upon the native vines; also, raspberries, etc., etc.

DES NONES. Color, clear yellow, with small brown dots; form, turbinated; size, 3; use, table; texture, white, buttery, melting, juicy; quality, 1; season, August to Sept.

REMARKS.—A pear of great excellence, combining the high flavor of the Seckel, with the delicious melting qualities of the Belle Lucrative, or its synonyme, Fondante d'Automne. In short, of very fine flavor and texture.

DELICES DE MONS. Color, lemon yellow; form, pyriform, uneven surface; size, 1; use, table; texture, melting, vinous, juicy; quality, 2; season, September.

REMARKS.—Foreign.

DE SORLUS. Color, yellowish, grayish, white dots; form, turbinate; size, 1; use, dessert; texture, white, melting, juicy; quality, 2; season, September to Nov.

Remarks.—New. Tree vigorous. Described favorably in *Hovey's Magazine.*

Dillen. Color, greenish yellow; form, obovate, obtuse pyriform; size, 1; use, table; season, September to Nov.
Remarks.—Tree not vigorous. Described in *Hovey's Mag.*

DILLER. Color, golden yellow, sprinkled with russet dots; form, obtuse, or one-sided; size, 2; use, table; texture, white, buttery, sugary; quality, 1; season, Aug.
Remarks.—Among the best. Its growth resembles the Bloodgood. Flesh, yellowish white, buttery, with rich, sugary, luscious flavor. A little gritty at the core. Seeds long, black, and pointed.

Dix. Color, yellow, with russet spots; form, oblong pyriform; size, 1; use, table; texture, melting; quality, 1; season, September to November.
Remarks.—First bore fruit in 1826, in the garden of Madame Dix, Boston. It is very hardy. It does not bear until it has reached some considerable size, different in that respect from the Bartlett and some others. Produces abundantly. Very free from disease. It deserves the attention of all cultivators, as it is of high excellence. Tree thorny. Very good, but long coming into bearing. Bears all at the top of the tree. Those pears on the quince that are the longest coming into bearing, are the best. Those well adapted to this stock come quite soon enough for the benefit of, and enduring effect on, the tree. The only difficulty is, that most of them are too early productive, and exhaust the vigor of the plant too much. In cultivating the pear on the quince, it is very desirable to know what kinds are the best for that purpose. There is a great deal involved in their particular adaptations for that purpose.

DOYENNE D'ALENCON. Form, obovate pyriform; size, 2; use, table; texture, buttery; quality, 1; season, November and December.

REMARKS.—Tree, fine, handsome grower. Good on quince or pear. A good bearer. A fine Winter fruit.

DOYENNE GRIS L'HIVER NOUVEAU, or *Winter Gray Doyenne*. Color, pale, dull yellow; form, obovate, obtuse pyriform; size, 3; use, table; texture, juicy; quality, 2 to 3; season, long keeper, to April.

REMARKS.—Flesh, tinged with orange, coarse-grained, but melting and juicy. Flavor, sprightly, vinous, good; slightly astringent near the skin. Ripens readily, in due time. It is recommended by the author, from experience, not to force the ripening of pears out of their natural season; but to mature them, gradually and evenly, in rather a low temperature, giving sufficient air when the weather is favorable. The air should be a medium, between dryness and a slight natural moisture. Not in extremes of either. The Gray Doyenne is described by Col. Wilder, in the *Horticulturist*, first established by Downing, and which leading work has also been well conducted ever since, by good theoretical, as well as eminently practical men. It is now in very good hands, and has, as it deserves, a large circulation, and still increasing, as may be expected from the gloriously growing interest in Horticulture, and Agriculture, in all parts of the Union. Men are beginning to get their eyes open, at length, to their highest welfare, happiness, and wealth. Fruit should comprise one-third of the human diet, at least.

DOYENNE D'ETE, or *Summer Butter*, of Indiana. Color, yellowish white; form, roundish, obtuse pyriform; size, 3; texture, juicy; quality, 2; season, August, and sometimes July.

REMARKS.—"Very good."—*J. B. Eaton*. Is a good early pear, but not very large. Tree grows slowly.

BUFFALO.

REMARKS.—Foreign. Tree, moderately vigorous. Early and abundant bearer. Flesh, a little coarse, but buttery, juicy, sugary, and sprightly.

DOYENNE PANACHEE. Color, yellowish, green, and red, in stripes; form, regular; use, table; texture, juicy; quality, 2 to 3; season, September and October.

REMARKS.—Singular in appearance. Not of high flavor.

DOYENNE DE CORNICE. Color, greenish yellow; form, turbinated; size, 1; use, table; texture, juicy; quality, 2; season, November and December.

REMARKS.—Flesh, melting, buttery, juicy, sugary, agreeably perfumed. Very delicious.

DOYENNE ROSE. Form, oblate, ovate pyriform; size, 1; texture, white, crisp, juicy; quality, 2; season, October.

REMARKS.—Foreign.

DOYENNE SIEULLE. Form, roundish obovate; size, 1; use, table; texture, melting, sugary; quality, 1; season, September and October.

REMARKS.—Foreign. Requires high culture.

DOYENNE ROBIN. Form, bergamot shaped; size, 1; use, table; texture, melting; quality, 1; season, September.

REMARKS.—Foreign. "Very fine. New, good, nearly first-rate."—*Fruit Committee, at Cincinnati Horticultural Exhibition*, 1855. Excellent here on quince.

Doyenne Santellette. Color, dull yellow, gray russet dots; form, roundish, pyriform; size, 1; texture, white, melting, vinous, juicy; quality, 1; season, September.

REMARKS.—Foreign. An old variety. Little known. Tree vigorous.

DOYENNE GOUBALT. Color, dull, pale yellow; form, obovate, acute pyriform; size, 2 to 1; use, table; texture, good; quality, 2.

REMARKS.—Foreign; great for orcharding on pear roots.

DOYENNE DE FAIS. Color, yellow; form, roundish; size, 1; use, table; texture, juicy, rich; quality, 2; season, September and October.

REMARKS.—Tree, robust, and productive. Described by Thorpe, Smith & Co., Syracuse, N. Y.

DUCHESS D'ANGOULEME. Color, dull green and yellow; form, oblate, ovate pyriform; size, 1 (often monstrous); use, table; texture, buttery, juicy; quality, 1; season, September and October.

REMARKS.—Magnificent. Sold for eight dollars per bushel by Professor Mapes and others, in New York City. Sometimes sold for seventy-five cents each in Philadelphia. Blooms too soon. Does not set so well with us—often very badly. Dr. Warder complains that he knows that six pears only set on fifty trees, three years planted, after having been white with blossoms, on quince stocks, on which it grows best. Valuable for market when it succeeds; some years free from frost, or some other cause affecting its bearing. High culture very desirable for it. Does well on the quince, and quince only, when it has a soil and climate to suit it. It always gets caught by the frost. We think it will not do here. The Duchess d'Angouleme should be cultivated only on the quince, and no where else. It is rather a shy bearer here, which, if it is not in extreme, is rather a good quality, as most pears are allowed to produce too largely on quince stocks, which makes them too short lived.

Duchesse de Mars. Color, yellowish russet; form, oblate obovate; size, 3; use, table; texture, melting, juicy, aromatic; quality, 2; season, September and October.

Remarks.—Best grown on quince, and succeeds well on it.

Duchesse de Berri. Color, pale yellow, with russet spots; form, round obovate; size, 2; use, table; texture, melting, juicy, sugary; quality, 2; season, August and September.

Remarks.—Tree, moderately vigorous.

DUNDAS, or *Parmentier*. Color, yellow, greenish black spots; form, obovate ovate; size, 2; use, table; texture, melting, aromatic; quality, 1.

Remarks.—Foreign—English. Rather liable to drop before matured.

DUNMORE. Color, green, red and russet; form, oblate obovate; size, 1; use, dessert; texture, white, buttery, melting; quality, 1; season, September and October.

Remarks.—A good grower and hardy bearer on pear stocks. Requires rich, high culture.

Early Butter of Cincinnati. Color, greenish yellow, dark greenish specks, sometimes a little of bronzy red in sun; form, oblong ovate, diminishing a good deal to the stem; size, 3; use, dessert; texture, white, buttery, juicy; quality, 2; season, last of July generally.

Remarks.—This is distinct from Dearborn's Seedling.

We know of no other description elsewhere. "Very good."—*Fruit Committee of Cincinnati Horticultural Society.*

Early Sugar. Color, green to yellow; form, pyriform; size, 3; use, table; texture, white, very saccharine or sugary, a little gritty; quality, 3; season, July.

Remarks.—Only valued for its earliness.

Early Butter of Indiana. Color, whitish yellow; form, obtuse pyriform; size, 2; use, table; texture, white, sugary, buttery, juicy; quality, 2; season, July.

Remarks.—"Not identical with any I have seen elsewhere."—*A. H. Ernst.* Tree, a long time coming into bearing.—There is a fruit near Newport, Kentucky, a seedling, which is a pretty good small early pear, generally ripe in July, which, just at a particular period in its maturity, and just before it rots or mushes in the core, is a pretty passable Summer fruit, although somewhat harsh and a little astringent at most points of its ripening. It bears well and is a healthy tree. It sells at about three dollars per bushel, chiefly on account of its earliness. One year part of the crop was ripe before the 4th of July.

Early Bergamot. Unworthy—very poor.

Early Catherine, or Early Rousselet. Unprofitable—poor.

EASTER BEURRE, with several synonymes, as *Beurre Gris d'Hiver Nouveau, Pater Noster, Doyenne de Printemps, etc., etc., etc.* Color, yellowish green, with russet spots; form, globular, obtuse pyriform; size, 1; use, table; texture, juicy, melting; quality, 1; season, December to March.

Remarks.—Requires a rich, warm soil, and some care in ripening, one of the arts not generally well understood, and owing to the want of good houses for preservation on good principles, similar to Schooley's, of Cincinnati, for instance, not successfully pursued. Col. Wilder, the best authority on this subject, speaks, from experience, very highly of Schooley's fruit houses and plans of saving fruit for a long period beyond their season. This fine pear was exhibited before the Cincinnati Horticultural Society in the Winter of 1855. Fruit Committee agreed it was "a first-rate Winter pear, and would keep a long time under proper and favorable circumstances." One of the best keeping table pears; first-rate in March.

EASTER BERGAMOT.

REMARKS.—Smaller, and not so good as Easter Beurre.

EASTER BEURRE.

REMARKS.—Rarely ripens. When it does it is excellent. Keeps well. See Beurre Easter.

ELIZABETH (Edwards). Color, lemon yellow; form, roundish, obtuse pyriform; size, 2; use, table; texture, white, crisp, melting, juicy, vinous; quality, 2; season, September and October. American.

ELIZABETH (Mannings). Color, lemon yellow; form, obovate, roundish; size, 3; use, table; texture, melting, sugary, juicy; quality, 1; season, August.

REMARKS.—Of foreign origin.

EDWARDS (William). Unprofitable.

EDWARDS (Henrietta). Color, dull yellow, spots of crimson in sun, russet round the stem; form, obovate, obtuse pyriform; size, 2; use, table; texture, rather coarse, but melting and juicy; quality, 2; season, August.

REMARKS.—Tree hardy, productive, and vigorous.

ENFANT PRODIGE. Color, russet or gray; size, 2; use, cooking.

REMARKS.—Flesh, greenish white; form, breaking, juice abundant, acidulous, but sugary, and very good.

EPARGNE. Color, green, with grayish spots of fawn color; form, long; size, 2; use, table; texture, buttery; quality, 2; season, August.

EPINE D'ETE, or *Winter Thorn.* Color, yellowish green; form, round at blossom end; size, 1; use, table and kitchen; texture, melting, tender; season, November to January.

REMARKS.—An agreeable flavor.

Eyewood. Color, greenish yellow; form, obovate; size, 3; use, table; texture, crisp, melting, juicy; season, September.

REMARKS.—Foreign. Tree vigorous, branches strong.

FIGUE. Color, green, with yellow; form, oblate pyriform; size, 2; use, table; texture, white, juicy, melting, aromatic; quality, 1; season, October.

REMARKS.—Tree vigorous, hardy. Productive on pear or quince.

FIGUE DE NAPLES. Color, pale greenish yellow; form, oblate, pyriform; size, 2; use, table; texture, juicy, sweet; quality, 2; season, October.

REMARKS.—This pear is the Beurre Bronzee of some. Tree vigorous and productive.

FIN OR D'ETE, or *Fine Gold of Summer.* Form, round; size, 2; use, table; texture, melting, juicy; quality. 2; season, July.

REMARKS.—Fine and beautiful. The skin is a little

rough; of a rich yellow on one side, and on the other a brilliant red, dotted with yellow. Growth of the tree vigorous, with long hanging limbs.

FLEMISH BEAUTY, or *Belle de Flandres* (which see); a beautiful, productive, and early bearer; not, however, equal to some. Great for market for short distances, with great care. Very tender; does not keep long. Must be gathered, like most other pears, a little before ripe, and while it is firm. This applies to the Bartlett, Seckel, Swan's Orange, Fondante d'Automne, etc., etc.

Flemish Bon Chretien. Unworthy of notice, or cultivation.

FONDANTE DE MILLET.

REMARKS.—Perhaps "Beurre de Millet;" but, however, it is "very good."—*Exhibition of the Cincinnati Horticultural Society*, 1855.

FONDANTE VAN MONS. Color, greenish yellow; form, roundish, obovate; size, 2; use, table; texture, white and melting, buttery, sugary; quality, 1; season, September.

REMARKS.—Foreign. Tree, good grower, and productive.

FONDANTE D'AUTOMNE, or *Belle Lucrative.* Color, pale yellowish green; form, pyriform; size, 2; use, table; texture, juicy, melting, aromatic.

REMARKS.—Extraordinarily fine. Exhibited by several, August, 1855, at the Horticultural Rooms. The Fruit Committee consider it "of the very finest pear flavor." "Excellently well adapted for the locality of Cincinnati." —*F. G. Cary, in Western Horticulturist and Cincinnatus.*

Fondante du Bois. Unprofitable; poor.

FONDANTE DE MALINES. Color, pale yellowish lemon; form, round obovate; size, 2; use, table; texture, juicy; quality, 2; season, September.

REMARKS.—Flesh white, buttery, melting; a little granulous near the core.

FONDANTE DE CHARNEUSE. Color, dull yellowish green; form, obtuse pyriform, irregular; size, 1; use, table; texture, melting, juicy, astringent.

REMARKS.—New. Described in *Horticulturist.*

Formes de Delices. Unworthy; below notice here.

FORELLE, or *Trout.* Color, green, striped yellow and red; form, pyriform; size, 2; use, table; quality, melting, juicy, vinous; season, September and October.

REMARKS.—Pretty, but not superior. Deserves some attention.

FORTUNEE, *or Episcopal, etc.* Color, grayish yellow; form, roundish; size, 3; use, table; texture, rich, melting, juicy, tender; quality, 1; season, October.

REMARKS.—"Very fine and good." *Horticultural Exhibition,* 1855. We do not know how Elliott placed this among the unworthy. It must have been from some unfortunate specimen he had. Such things will happen sometimes, even among the most careful and skillful.

FREDERIKA BREMER. Form, round obovate; size, 1; use, table; texture, brittle; quality, 2; season, September and October.

REMARKS.—Flesh, white, free from grit, melting, fine, buttery, sugary, slightly acid. Described by J. C. Hastings in *Horticulturist.* "A good pear."—*Fruit Committee of Cincinnati Hort. Society, Horticultural Exhibition,* **1855.**

FREDERICK DE WURTEMBURG. Color, dull yellow; form, angular pyriform; size, 1; use, table; texture, white, juicy, melting; quality, 2; season, August and September.

REMARKS.—Tree early and productive. Decays at core. Sometimes very large and handsome, but uncertain. A less vigorous tree than desirable.

Franc Real d'Hiver.

REMARKS.—Poor, undeserving attention.

FRANGIPANE. Color, yellow; form, long; size, 2; use, table; texture, melting; quality, 2; season, September.

REMARKS.—Perfumed.

FULTON. Color, reddish, or dark cinnamon russet; form, flattened round; size, 2 to 3; use, table; texture, buttery; quality, 1; season, September and October.

REMARKS.—From Maine. Very fine. Tree very hardy, and abundant bearer. Well suited for standard orcharding in the West.

GANSEL'S BERGAMOT. Color, yellow, with brownish russet; form, roundish obovate; size, 2; use, table; texture, crisp, sugary; season, September.

REMARKS.—Requires a warm, rich soil. Succeeds well on quince. Fruit a little gritty at core, but of extraordinary high flavor. Does not grow so large here as in England, owing, probably, to too warm a climate, and, therefore, never equal to its English, or foreign, reputation. Tree not very vigorous here.

Gendesheim. REMARKS.—Undeserving attention.

Gideon Paridante. Color, yellowish gray, brown in sun; form, obtuse pyriform; size, 2; use, table; texture, melting, juicy, sugary; quality, 2; season, August and September.

REMARKS.—A respectable fruit, though not first-rate.

GILE-O-GILE, *or Garde d'Ecosse.* Color, russet, with a reddish russet cheek; form, roundish; size, 1; use, cooking and preserving; texture, firm and crisp; quality, 1, for baking, etc.; season, October to January.

REMARKS.—Esteemed highly for preserving. Bears large crops. We have already noticed the necessity of watering fruit trees, and particularly recommend it to pear trees that have large crops on them, as it will prevent them falling off, and assist greatly in their size, for the crops of standard trees can not well be thinned. Wet Summers alone have convinced the author how much these trees are assisted in their crops by plenty of moisture. And the experience of very dry Summers has proved the reverse.

GLOUT MORCEAU. Color, pale yellowish green; form, obovate, obtuse pyriform, often angular; size, 2; use, table; texture, juicy, melting; quality, 1; season, December to February, or later.

REMARKS.—Flemish, and one of the best Winter fruits. Honied. Does not bear well when young. After ten years, an abundant bearer. One of the best and hardiest, and fine. The Glout Morceau is the best Winter pear in this vicinity, and is almost the only one that ripens well here, without extraordinary pains being taken for that purpose in a particularly careful manner. Indeed we do not seem to have found out a good method of preserving and ripening Winter pears, suitable for our climate, yet. The Glout Morceau, and Vicar of Winkfield or Clion, and Winter Nelis, are about our best Winter pears. As a general thing, Winter varieties are difficult to ripen. The chief reason seems to be, that the leaves fall too soon. Nature can not perform her functions, and, therefore, the fruit is deficient in those peculiar qualities for ripening, which are necessary for it. Hot, dry weather, of which we often have so much, deprives the tree prematurely of its leaves. It is not so frequent in the East as with us; we having a richer soil also. The Glout Morceau is one of the latest to shed its leaves. Those kinds which hold their leaves longest, are the best trees to bear Winter fruit. Winter Nelis is likely to be a good early bearer here. Is a tolerably good grower, although it is hard to make a handsome tree of it. The Glout Morceau bears well on the quince, but ripens with some, rather poorly. Others speak well of it for ripening, compared with other Winter pears generally.

GOLDEN BEURRE OF BILBOA. Color, rich yellow, with russet dots; form, obovate; size, 2; use, dessert, and beautiful; texture, juicy, melting, sweetish; quality, 1; season, August and September.

REMARKS.—Spanish origin. Tree hardy. Requires a rich, strong, heavy soil. Exhibited by Wm. Heaver, and Mr. Hatch (those of the latter very finely colored, as is usual with this gentleman's fruit, from some cause; perhaps, sunny exposures and superior cultivation), August, 1855. Decision of the Fruit Committee, "Juicy, sweet, buttery, melting, and good, but not the highest flavor." Very fine, but not superior, if equal, to Belle Lucrative, or Fondante d' Autumne. It is a rich looking and pretty good pear.

GRAND SOLEIL. Color, orange yellow; form, roundish; size, 2; use, table; texture, white, crisp, buttery, melting; quality, 2; season, October.

REMARKS.—Tree vigorous, well formed, productive.

GREEN CATHERINE, or *Rousselet.* See *Rousselet.*

GREEN MOUNTAIN BOY. Color, golden yellow; form, round obovate, irregular; size, 2 to 1; use, table; texture, melting, juicy, sugary; quality, 2; season, September.

REMARKS.—"Native American."

GREEN CHISEL, see *Madeleine.*

GRAY DOYENNE, with many synonymes. Color, cinnamon russet; form, roundish, obovate; size, 2; use, table; texture, melting, juicy, buttery; quality, 1; season, September.

REMARKS.—Distinct from Boussock and Surpasse Vergalieu. A good bearer on pear or quince. A hardy tree. A small, very early green pear, never turning yellow, rotting rapidly—a very poor fruit, nothing like the Madeleine, which is one of our best Summer fruits.

Green Pear of Yair, or Grand Monarque. A poor affair.

GRISE BONNE, or *Good Gray Pear.* Color, gray, with black spots; form, regular; size, 2; use, table; texture, juicy; quality, 2; season, August.

REMARKS.—Flesh, large grained, but juicy.

Groom's Princess Royal. Color, greenish brown; form, roundish; size, 2; use, table; quality, unknown in America, but 1 in England; season, Winter.

REMARKS.—Not yet tested here. Raised by Mr. Groom, the famous tulip grower, near London, who has earned a good name as a great fruit and flower grower.

Grosse Calebasse. Color, greenish yellow; form, bell; size, 1; use, kitchen; texture, crisp, dry; quality, 3.

REMARKS.—Considered unworthy of cultivation by A. H. Ernst, who has paid great attention to pears, and has had a larger variety growing at Spring Garden Nursery, than any other person in the neighborhood of Cincinnati.

HAMPDEN'S BERGAMOT. Color, green at first, yellow at maturity, with small dots, and a few greenish specks, sometimes, in the shade; size, 1; use, table; quality, 2; season, August, sometimes July.

REMARKS.—A beautiful fruit; approaches, at times, near to the attraction of the Bartlett, but inferior in flavor to it, and not near so buttery, but breaking in texture.

HADDINGTON. Color, greenish yellow; form, ovate, obovate pyriform; size, 1; use, table; texture, juicy, aromatic; quality, 2; season, Winter.

REMARKS.—From the seed of the Pound Pear. Tree, vigorous and productive.

HACON'S INCOMPARABLE. Color, dull yellowish green; form, round, obtuse pyriform; size, 2 to 1; use, table; texture, buttery, melting, sugary; quality, 2; season, September and October.

REMARKS.—Hardy, productive; deserving of attention.

HAGERMAN. Color, yellow; form, roundish; size, 3; use, dessert; texture, juicy, sugary; quality, 2; season, August and September.

REMARKS.—From Flushing, New York.

HANOVER. Color, green; form, round obovate; size, 3 to 2; use, table; texture, melting; quality, 2.

REMARKS.—Hanover Furnace, New Jersey.

HANNERS. Color, yellowish green; form, oblate; size,

2; use, table; texture, melting, juicy, vinous; quality, 2; season, August and September. Origin uncertain.

HARRISON'S LARGE FALL. Form, flat at blossom end; size, 2; use, baking; texture, crisp; season, September.

REMARKS.—Fall Baking of Coxe.

HARVARD. Form, oblate pyriform; size, 2; use, table; texture, white, juicy; quality, 2; season, September.

REMARKS.—From Cambridge, Mass. Tree hardy, vigorous, upright, productive. Fruit liable to decay at core.

Hativeau.

REMARKS.—A very small pear, pointed toward the stem; the blossom end flat; the skin is clear yellow, the flesh is of a yellowish cast, somewhat spicy, but without much juice or flavor. It is a very great bearer; the time of ripening from the middle to the end of July. Authority, Coxe, of Burlington, New Jersey, an eminent practical Pomologist.

Hayel, of *Hessell.* Color, greenish yellow; size, 2; use, kitchen; quality, 3; season, September.

REMARKS.—"Flesh sweet; rough, and astringent, however. Unworthy."—*A. H. Ernst.*

HENRIETTA. Color, dull yellow; form, obovate, obtuse pyriform; size, 2; use, table; texture, crisp, melting, juicy; quality, 2; season, August.

REMARKS.—Of American origin—New Haven, Conn. Tree vigorous, hardy, productive.

HENKEL. Color, dull yellow, russety; form, obovate; size, 1; use, table; texture, crisp, juicy, vinous; quality, 2; season, September.

REMARKS.—Foreign. Tree fine grower; early and productive bearer on pear roots. Valuable for orcharding.

Hericart. Color, pale greenish yellow; form, obtuse pyriform; size, 1; use, table; texture, crisp, juicy; season, August and September.

Hessel. Unworthy of culture.

HEATHÇOT (Gore's). Color, greenish yellow, skin rough; form, obovate, rounded; size, 1 to 2; use, table; texture, melting, buttery, juicy; season, September.

REMARKS.—From Waltham, Mass. Not much known, but very deserving. Tree hardy, branches slender. A very fine pear. Tree thrifty, pyriform. Holds its leaves well. Productive when old enough, and very good. Medial, juicy, but not rich. We know of no fruit that varies in quality more than the pear; for, while some of the kinds are so unpalatable as not only to be refused by the swine, but even rejected by ravenous boys, others are of so delicious a flavor that we see most of the Autumnal fruits give place to them. A bad pear is injurious to health, and brings a poor price in the market, while a rich, melting pear, is eagerly sought after at a large price; and we never recollect of any person being injured by eating such fruit. Those, therefore, who have discovered their trees to produce a worthless fruit, should lose no time in grafting them with a known good variety, that both themselves and future generations may be benefited.

HOLLAND GREEN, or *Holland Table.* Color, green, with small spots; form, irregular; size, 1 to 2; use, table; texture, juicy; season, August and September.

REMARKS. — Tree a strong and vigorous, with long branches; foliage luxuriant. It is a great and uniform bearer. Imported from Holland by Wm. Clifton, of Philadelphia.

Holland Bergamotte. Color, brown, with spots; form, flat at crown; size, 2; use, table; texture, melting; quality, 2; season, Winter. "Good and juicy."—*Horticultural Exhibition*, 1855, *by the Fruit Committee.*

HONEY. Color, rich golden russet; form, round, tapering; size, 3; use table; quality, 1; season, August.

Remarks.—Delicious. Season a little before the Bartlett.

HOWELL. Form, obtuse pyriform; size, 1; use, table; texture, juicy; quality, 1; season, September and October.

Remarks.—One of the most beautiful in cultivation. Flesh, melting and juicy; flavor, rich, slightly acidulous, with a delicate aroma.—*Wilder, in Horticulturist.*

Huguenot. Unworthy.

Hull. Color, yellowish green, with some dull red and russet; form, obovate; size, 2; texture, white, a little coarse, gritty at core; quality, 2; season, September.

Remarks.—Tree upright and strong.

Inconnue Van Mons. Color, pale green; form, oblate pyriform; size, 2; use, table; texture, melting; quality, 2; season, Winter.

Remarks.—Foreign. Best on Quince. Flesh, melting, buttery, and fine; flavor, pleasant, good, resembling a little the Glout Morceau.—*Col. Wilder, Horticulturist.*

Jaminette. Color, green, with russet dots; form, obovate, obtuse, pyriform; size, 1; use, table; texture, juicy, sugary; quality, 2; season, November and December.

REMARKS.—Of foreign origin. Tree vigorous. Not an early bearer. Fruit a little gritty at core.

Jargonelle English. Color, light green; form, long neck; size, 2; use, table; texture, juicy; quality, 3; season, July and August.

REMARKS. — Should be picked a little time before fully ripe, or it will soon rot at core on the tree, and fall. Upon the whole, it may be called unworthy. Exhibited by T. V. Petticolas, and the author, at the same time, July, 1855. Fruit Committee reported, "This fruit is not much admired; though, when eaten just in the nick of time before it is too ripe, tolerably good. It soon rots at the core." If picked before ripe, and kept a day or two, is not so very bad. Bears well on the quince.

Jargonelle French. Of still less value.

JALOUSIE. Color, of the deepest russet, ruddy in the sun, and curiously marked with lighter colored specks, a little raised; form, varying from roundish to obovate, and more frequently pyriform; size, 1 to 2; use, dessert; quality, 2; season, September.

REMARKS.—It is only of second-rate flavor, and soon rots at the core. An unique-looking old French pear. Exhibited by A. H. Ernst, and reported by Fruit Committee, "Pretty good."

JALOUSIE DE FONTENAY VENDEE. Color, dull yellow and green, with russet patches and dots, and with a red cheek; form, turbinate, or obtuse pyriform; size, 2; texture, white, buttery, melting, with a rich flavored juice; quality, 1; season, September and October. It gives general satisfaction. Fine, grown on quince; inclined to overbear. "Very good."—*Fruit Committee Report*, 1855.

Jaminette. Only "good." Fruit above medium.

JEAN DE WITTE. Color, pale yellow; form, obovate; size, 1 to 2; use, table; texture, melting, juicy, vinous; quality, 1; season, October to December.

REMARKS. — One of Van Mons. Tree of moderate growth.

JERSEY GRATIOLE. Color, greenish yellow; form, roundish oblate, obovate, angular; size, 1; use, table; texture, white, crisp, sugary, vinous; quality, 2; season, Sept.

REMARKS.—Tree moderately vigorous. Great bearer. Succeeds on quince. Described and commended in *Hovey's Magazine.*

JOHONNOT. Size, 2 to 3; use, table; texture, crisp, a little melting, juicy; quality, 2; season, September.

REMARKS.—An Eastern fruit, from Massachusetts. Tree moderately healthy. Productive in the Western soils.

Josephine de Malines. Color, yellow, russety; form, roundish obovate; size, 2; use, table; texture, juicy, melting, vinous; quality, 1; season, November and December.

REMARKS.—Not an early bearer. Best on pear. To render trees more fruitful, the author has witnessed the advantage of taking off a ring of bark from the branches of unfruitful pear trees, and he has no hesitation in advising the practice, as it has not been found to injure the trees so operated upon, on which it has been repeated for several successive years. But it should be performed on branches that are not exposed to the full rays of our fervent sun. Many persons object to the experiment, fearing to injure the branches; but what is the value of unproductive limbs. The author has seen, in the month of May, a ring taken off of the bark of two principal branches, the one extending east and the other west. In the following Spring, these two branches were covered with flowers, although no other part of the tree gave out a single blossom. The author saw this tree in the Fall, and counted thirty-nine fine grown pears on one of the ringed branches. Some gardeners have regarded the plan as being unnatural, as if it was more unnatural to make a tree fruitful by checking the circulation of the sap, than to make it productive by grafting.

JULIENNE, *or L'Archduc D'Ete of Coxe*, sometimes called *Summer Beurre.* Color, brownish yellow; form, roundish; size, 2; use, dessert; texture, juicy; quality, 1; season, August, sometimes July.

REMARKS.—Larger than Bloodgood, or Dearborn. Bears in three years on its own stock. Exhibited at the Cincinnati Horticultural Rooms by many members during the season. Fruit Committee report it "A first-rate Summer fruit." Very good, if house ripened, equal to Doyenne Gris, the Brown Buerre of Coxe. One of the best early pears. Bears young and abundantly, on alternate years. The tree is of singular growth, the branches long and bending, with large swellings at the extremities. The wood of a lively yellow brown. It is sometimes several weeks in perfection. It bears young and most abundantly. It is called by some the butter pear of Summer. It should be gathered before fully ripe, and kept a few days in the house, as before observed.

KINGSESSING. Form, obovate, sometimes obtuse; size, 1; use, table; texture, buttery; quality, 1; season, August.

REMARKS.—From near Philadelphia. Not an early bearer. Requires double working on quince.

King Edward's. Color, yellow; form, pyriform; size, 1; use, table; texture, buttery; quality, 2; season, October.

Kirtland Pear. Color, crimson russet; form, globular, ovate; size, 2; use, table; texture, juicy; quality, 1; season, September.

Remarks.—Raised by Dr. Kirtland, of Cleveland, Ohio, a good pomologist, a man of very scientific attainments, and of versatile talents. Texture, fine, melting, juicy and rich. Flavor, aromatic, sweet, and in the highest degree delicious. Tree, hardy and productive. Obtained from the seed of Seckel. Dr. Kirtland has been most successful in raising very valuable and delicious cherries from the seed (which will be noticed in their proper place), as well as some other fine seedling fruits. He is always indefatigable in the pursuits of horticulture and science.

Knight's Monarch. Color, yellowish brown; form, obovate oblate; size, 1; use, table; texture, melting, juicy; quality, 1; season, December.

Remarks.—Of a somewhat musky flavor. Requires high cultivation. Rather similar in shape to the Flemish Beauty, or Belle de Flandres, but a little larger. This is the fruit, some spurious trees of which Mr. Knight having been the means of distributing, he stated that he would have rather lost ten thousand pounds than have done so.

Knight's Seedling. Color, yellowish green, with a brownish red cheek; size, 2; use, table; texture, juicy, sweet; quality, 1; season, September.

Remarks.—From Cranstown, Rhode Island. "A beautiful and excellent fruit."—*A. H. Ernst.* Tree vigorous and early productive.

La Juive. Color, yellow; form, pyriform; size, 1; use, table; texture, buttery, juicy; quality, 2; season, September.

Remarks.—Foreign. Described in *Hovey's Magazine.*

Large Summer Bergamot, see *Summer Bergamot.*

Las Canas. Color, pale yellow; form, pyriform; size, 2; use, table; texture, juicy, sugary; quality, 1; season, September.

Remarks.—Of French origin. Tree, vigorous, and an early bearer.

Laherard. Color, lemon yellow; form, obtuse pyriform; size, 2; use, table; texture, juicy, white; quality, 1; season, September.

Remarks.—Succeeds on quince. Described by Wilder in *Horticulturist.*

LAWRENCE. Color, dull pale yellow; form, obovate, obtuse at stem; size, 2; use, table; texture, juicy, melting; quality, 1; season, October to February.

Remarks.—Recommended by Ohio Pomological Society for general cultivation. Very good. Tree rather thorny. S. B. Parsons says it succeeds well on the quince. A little gritty at core. Slightly sugary. A great and early bearer. A good keeper—till Christmas. Retains its fine juicy quality, and does not shrivel in ripening.

L'Echasserie. Color, yellow; form, oval; size, 2; use, table; texture, melting, vinous; quality, 2; season, November to January.

Remarks.—A pretty good pear, although rather too small—which is a fault, particularly for market. We would not recommend it.

LE CURE, see *Vicar of Winkfield, Monsieur Le Cure, or Clion.* A pear very profitable for market—productive and early.

Little Muscat. Color, yellow, with a dull red cheek; form, turbinate; size, 4 (very small); use, table; texture, breaking, sweet, with a slight musk flavor; quality, 2 to 3; season, July, sometimes even June.

REMARKS.—French. It is the earliest of pears, and that is its chief merit. It bears most enormously, in clusters. It is only passably good. Can be sold by the quart at eight or nine dollars per bushel. It is very common to see it on the same stall with cherries. Trees very handsome, of fine pyramidal growth, and hardy.

LOCKE. Color, yellowish green; form, round obovate; size, 2; use, table; texture, juicy, vinous; quality, 2; season, October.

REMARKS.—American—West Cambridge, Mass.

LONG GREEN. Color, green, dark gray spots; form, oblate, ovate pyriform; size, 2; use, table; texture, melting; quality, 1; season, September and October.

REMARKS.—An old foreign variety; always very good, and an abundant bearer.

LOUISE BONNE DE JERSEY. Color, smooth and glossy, green in shade, brownish red in sun, dotted with gray dots; form, oblong pyriform; size, 1; texture, white, juicy, melting, and delicious; quality, 1; season, September and October.

REMARKS.—One of the best and most productive large pears. Best on quince. From the Island of Jersey. Tree hardy, very productive, and very early in bearing. Fruit rather variable in quality. Shoots vigorous. Fruit of nearly the first order of merit; very good, but does not grow large enough here; still it is a good market fruit. Louise Bonne de Jersey just suits the quince, and bears moderately, and not too much to shorten a great deal the life of the tree.

LOUIS PHILLIPE.

REMARKS.—The tree came from England. It is a great bearer; not quite so large as the Bartlett, but nearly as good. Comes in later. Most of our best pears, except our native, are from England and France. The apples are more our own; nine-tenths of American apples are native; three-fourths of the pears we cultivate are foreign. American fruits will prove the hardiest for cultivation, even if the fruit from them is not so good. Pears are more of an exotic luxury with us than apples; they are altogether the more difficult for us in the West to raise than apples and other fruits. We can not expect but to have difficulties with pears on the quince, as we have had, and still, to some extent, suffer, and probably shall still be subject to, with standard pears. Our climate is more dry than those States on the Atlantic border. The pear delights in moisture.

LODGE. Color, yellow; form varies; size, 2; use, table; texture, juicy, melting, a little gritty; quality, 2; season, September.

REMARKS.—Native born—from Philadelphia. Requires a rich, warm, sandy soil.

MADELEINE, *Citron des Carmes, or Green Chisel.* Color, green; form, obovate oblate; size, 3; use, table; texture, buttery; quality, 1; season, July.

REMARKS.—A uniform bearer. The fruit is slightly acid, but delightful. One of the very best early pears;

productive; growth too upright. Taste, when not too ripe, sugary. This pear Mr. Prince calls Chaumontel. It does well either on pear or quince stocks. There is a striped Madeleine, Citron des Carmes Panachee (which see), which differs from the above in being striped with light yellow: inferior in flavor. The Madeleine is not very vigorous in growth, but sufficiently so to do well.

MANNING'S ELIZABETH. See Elizabeth.

Marie Louise. Color, dull green; form, oblate pyriform; size, 1; use, table; texture, buttery, melting, vinous, juicy; quality, 2 to 3; season, September and October.

REMARKS.—Fruit Committee reported this fruit rather inferior at the Horticultural Exhibition, 1855. It is only passable, and hardly that. Tree vigorous, hardy, and productive.

March Bergamotte. Color, greenish yellow; form, obovate; size, 2; use, table; texture, buttery, gritty at core; season, March.

REMARKS.—Poor. It is difficult to keep Winter pears in this climate, owing partly to its great irregularity, and partly to the want of preparation for the purpose; also, fruit in general is not sufficiently tenderly handled in the gathering. The slightest bruise is calculated to rupture the cells and break the fine tissue. It is not easy to maintain the right temperature. This can only be surely and effectually accomplished by building fruit houses for the express purpose. See Schooley's Patent.

MARTIN SEC. Color, yellow; form, obovate, angular; size; 2; use, table; texture, melting, juicy; quality, 2; season, Winter.

REMARKS.—A foreigner.

MARECHAL DE LA COUR. Color, yellowish green; form, oblate pyriform; size, 1; use, table; texture, melting, juicy, vinous; season, October.

REMARKS.—An adopted citizen. Tree vigorous. Described in *Hovey's Magazine.*

Marie Louise Nova. Unworthy.

MELLA DE WATERLOO. Color, greenish yellow; size, 1; use, table; texture, melting, rich, and juicy; quality, 1; season, October.

REMARKS.—"A new pear of great excellence."—*A. H. Ernst.*

MERRIAM. Color, yellow; form, round; size, 1; use, table; texture, crisp, juicy, sugary; quality, 2; season, September.

REMARKS.—From Roxbury, Massachusetts. Described, and favorably, in *Hovey's Magazine.*

MERVEILLE D'HIVER, or *The Wonder of Winter.* Color, dull green, with russet spots; form, irregular; size, 3 to 2; use, table; texture, melting, luscious; quality, 2; season, December.

REMARKS.—Destitute of beauty, but otherwise pretty good. Eye, very singular, frequently without any crown.

MESSIRE JEAN. Color, yellow, with some russet; form, round at blossom end, small at stem; size, 2; use, table; texture, juicy; quality, 1; season, October and November to December.

REMARKS.—Excellent Winter pear. In the possession of A. Worthington, who thinks well of it. Not very rich or high flavored. On young trees it sometimes grows large. Flesh, rather coarse, and sprightly.

MILLOT DE NANCY. Color, pale yellowish green; form, obovate pyriform; size, 2; use, table; texture, juicy, sugary; quality, 2; season, Winter.

REMARKS.—Tree, very vigorous, of rapid growth, and productive. Wilder, in *Horticulturist.*

MOON'S POUND. Color, lemon yellow; size, 2; use, kitchen; texture, melting, juicy; quality, 1; season, Dec.

REMARKS.—Exhibited by A. H. Ernst, at the rooms of the Horticultural Society, August 21, 1855. Fruit Committee report it, "A very good pear." "Very tender, and abounding in pleasant juice. May prove a fine fruit." —*A. H. Ernst.* Valuable for cooking, and a profitable orchard fruit.

Michaux. Unworthy.

MITCHELL'S RUSSET.

REMARKS.—A seedling from the Seckel; resembling it in every respect, but of larger size. Origin, Belleville, Illinois. *Fruits of Missouri*, by Thos. Allen, of St. Louis.

Moor Fowl's Egg.

REMARKS.—Unprofitable, and, therefore, unworthy.

MONSIEUR LE CURE. See *Vicar of Winkfield.*

MOLLETT'S GUERNSEY BEURRE, *or Chaumontelle.* Color, yellowish green, with dusky brown, some russet in the sun; size, 2; use, table; texture, yellowish, melting, vinous; season, October and November.

REMARKS.—"Very good."—*Elliott's American Fruit Grower's Guide.*

MOCCAS. Color, greenish yellow, brownish cheek in

sun, some russet specks; size, 2; use, table; texture, yellowish, melting, juicy; season, December.

Remarks.—Tree, very vigorous, growth very quick and prolific.

Muscat Allemande, *or German Muscat.* Color, green; form, blossom end wide and flat; size, 2; use, table; texture, rich and buttery; quality, 2; season, October.

Remarks.—Flesh, yellow. It is a good pear. From France.

Muscat Robert, or Little Muscat. See Little Muscat. A small pear, very early, and very poor when too ripe. If it is not too ripe, it is rather pleasant. It ripens from the 1st to the 10th of July.

MOYAMENSING (Smith's). Color, lemon yellow, with yellowish russet; form, round, obovate; size, 2; texture, melting, buttery; quality, 1; season, August.

Remarks.—A native of Pennsylvania. Tree, vigorous, wood yellowish brown, with light dots. A regular and abundant bearer.

Muscadine. Color, yellowish green; form, round obovate; size, 2; use, table; texture, white, buttery, melting; season, August.

Remarks.—Not prolific. A native American.

Musk, Summer Bon Chretien, *or Large Sugar.* See Summer Bon Chretien. "An old, and once good variety, but now not very worthy."—*A. H. Ernst.*

NAPOLEON. Color, greenish yellow; form, obtuse pyriform; size, 1; use, table; texture, buttery, juicy; quality, 1; season, September.

REMARKS.—Foreign. Very fine, large, juicy, vinous, hardy, and thrifty. Tree, vigorous. Fine on quince. Productive. Good for market in every respect. Deserves general cultivation. Bears young and regularly, and is good and large enough.

Naumkeag. Unworthy.

NIELL D'HIVER.
REMARKS.—"Pretty good, rather harsh, and rough."—*Fruit Committee Report, Horticultural Exhibition*, 1855.

Ne Plus Meuris. Color, dull yellowish brown; form, roundish; size, 2 to 3; use, table; texture, buttery, melting, with a sugary and very agreeable flavor; quality, 2; season, end of Fall into Winter.
REMARKS.—Keeps well, under proper care and advantages. Its appearance is unprepossessing, and uneven. Belgian.

Nouveau Poiteau. Color, pale green, dark green spots; form, obovate, obtuse pyriform; size, 1; use, table; texture, juicy, melting, rich, aromatic; quality, 2; season, October.
REMARKS.—"Flesh, melting and juicy; flavor, rich, sweet and delicious, with melon-like aroma. Medium fruit." Described by Colonel Wilder, who considers it "an acquisition." Said to be, by many, a good Winter pear.

ONONDAGO, see *Swan's Orange*, one of the very best. Deemed by Mr. A. H. Ernst very fine. Sometimes twelve inches in circumference.

Orange Musquee, or Musk Orange. Color, greenish yellow;

form, round; size, 2; use, table; texture, juicy; quality, 3; season, July, and sometimes August.

REMARKS.—Uncommonlv subject to rot.

OSWEGO BEURRE, see *Beurre Oswego*. Color, dull yellowish green, with marblings and patches of russet; form, ovate obovate, or obovate rounded; size, 2; use, table; texture, melting, juicy, sub-acid, sprightly; quality, 1; season, September.

REMARKS.—Bears early on quince or pear roots. American origin. "Rather acid."—*J. B. Eaton, Buffalo.*

OSBORN.

REMARKS.—Fruit small, ovate, pyriform, greenish yellow; stem about one inch long, set with two or more fleshy ridges at base. Flesh, white, tender, juicy, sweet, very little astringent. A very good Summer pear. Introduced first to notice here by A. H. Ernst.

OTT PEAR. Form, rounded; size, 3; use, table; texture, juicy; quality, 2; season, August.

REMARKS.—Flavor aromatic, rich and sweet. Described by Dr. Brinckle in *Horticulturist*, Philadelphia. Seedling from the Seckel, and hardy and healthy like the parent tree. Best, also, like it, grafted on the apple, leaving part of the branches of the stock or apple to grow. This plan gives great size and healthy development to the fruit, as advised by Mr. Sleath, formerly the very experienced and scientific pioneer gardener of Cincinnati and N. Longworth, and who did much here for horticulture, at an early day. By this mode the too strong tendency of the Seckel to grow to too much wood, is checked.

ORANGE BERGAMOT. Color, yellow; form, flat at blossom

end; size, 1; use, baking; texture, crisp; quality, 2; season, September.

REMARKS.—"Only fit for baking."—*A. H. Ernst.*

PANACHEE, see *Verte Longue.*

PADDINGTON, or *Easter Bergamot.* Color, green; form, round; size, 1; use, table; texture, sprightly, crisp, juicy, and melting at maturity; quality, 2; season, January to February.

REMARKS.—A little acid. It is inferior to the Easter Beurre for eating, but worthy of cultivation. Stem short and thick.

PASSE COLMAR, with about twenty synonymes. Color, yellowish green, becoming yellow at maturity, sprinkled with light brown russet; form, oblate, obovate, obtuse pyriform; size, 1; use, table; texture, buttery, sugary; quality, 2, sometimes 1; season, Winter.

REMARKS.—Uncertain in quality at the West. Slender wood, thrifty. Mr. Buchanan tells us to be patient with it, and we shall be repaid. There have been brought to the Horticultural Society, occasionally, a few good specimens. The tree requires high culture, severe pruning, and great care, being too luxuriant in growth.

PAYENCY. Color, dull pale green, with patches of russet, yellowish at maturity; form, pyriform; size, 2; use, table; texture, juicy, sugary, melting; quality, 2; season, September.

REMARKS.—Tree of vigorous growth, early bearer, and particularly profitable on the quince.

PETRE. Color, pale yellow; form, oblate, obovate

pyriform; size, 1 to 2; use, table; texture, melting, buttery, juicy; quality, 1; season, September.

REMARKS.—From Bartram's Botanic Garden, Philadelphia. Tree moderate grower.

PENDLETON. Color, greenish yellow; form, obovate, obtuse pyriform; size, 1; use, table; texture, tart, sugary; quality, 2; season, July.

REMARKS.—American—Connecticut. Described favorably in *Hovey's Magazine.*

PENNSYLVANIA. Color, yellowish green; form, obovate pyriform, irregular; size, 1; use, table; texture, crisp, buttery, juicy; quality, 2; season, September.

REMARKS.—American. Tree, vigorous, hardy and moderately productive. "A pretty good pear."—*Dr. Warder.*

Pitt's Prolific. Unworthy.

PLOMBGASTEL. Color, greenish yellow, red in sun; form, ovate pyriform; size, 1; use, table; texture, melting, juicy; quality, 2; season, August and September.

REMARKS.—Tree vigorous. Succeeds on quince.

POIRE D'ALBERT. Form, pyramidal, angular; size, 2; use, table; texture, white, crisp, melting, juicy; season, September.

REMARKS.—Tree, vigorous, and productive on pear. Described in *Hovey's Magazine.*

Poire de Jardin, or Garden Pear. Color, yellow; form, long and flat; size, 1; use, table; texture, juicy; quality, 2; season, October.

Pope's Scarlet Major. Unworthy; poor.

Pope's Quaker. Unprofitable; indifferent; poor.

POUND. Color, greenish brown; form, regular; size, 1; use, baking; texture, juicy; quality, 1; season, Winter; keeper, well and long.

REMARKS.—Weighs sometimes from twenty-six to twenty-eight ounces. Tree, very vigorous, and large, and hardy. "Productive and hardy, keeps well, but good only to sell."—*Fruit Committee Cincinnati Horticultural Society.*

Pound, Moon's. See Moon's Pound.

Princess of Orange. Not worth notice.

PRATT. Color, greenish yellow; form, obovate pyriform; size, 1; use, table; texture, melting, juicy; quality, 1; season, September.

REMARKS.—From Rhode Island. Tree, vigorous, and productive on pear.

PRINCE'S ST. GERMAIN. Color, green, and pale yellow, marbled; form, oblate, obovate pyriform; size, 2; use, table; texture, juicy; quality, 2; season, November to February.

REMARKS.—Fine, juicy, Christmas pear. From the seed of the old St. Germain, nearly fifty years since, at the Princes' Nurseries, Flushing, Long Island. Tree, hardy, thrifty, very productive. Fruit, keeps well, and requires no more care to ripen, than apples (a rare thing). Shoots, reddish brown.

PULSIFER. Color, yellow; form, pyriform, obtuse; size, 3; use, table; texture, white, melting, juicy; quality, 1; season, August.

REMARKS.—Native to the soil. Tree vigorous.

Queen of the Low Countries. Color, dull yellow; form, obovate, acute pyriform; size, 1; use, table; texture, buttery, melting, juicy; quality, 2; season, September.

Remarks.—Not a great bearer in the West.

Quilletette. Unworthy; poor.

Rallay. Color, yellow; form, globular, acute pyriform; size, 3; use, table; texture, juicy, gritty at core; quality, 2; season, Winter.

Remarks.—Ellwanger & Barry.

Raymond. Unworthy of notice.

Red Bergamot. Color, yellow, with much red to sun; form, round; size, 2; use, baking; texture, juicy; quality, 2; season, August.

Remarks.—Tree, of vigorous growth, and large size. It does not bear while young, but when more advanced is a great bearer. "Only for baking."—*A. H. Ernst.*

Reading. Color, greenish yellow, with dots; form, pyriform, tapering to crown; size, 2 to 1; use, table; texture, sprightly, vinous; quality, 2; season, January to March (maturity).

Remarks. — "Good." — *Downing's Horticulturist,* New York. From Philadelphia. Young shoots, tender, yellowish olive; old wood, gray olive.

Reine Caroline. Color, yellowish green, becoming yellow at maturity, with a rich, brownish red cheek; form, narrow pyriform; size, 2; use, cooking only; texture, dry and poor; quality, 3; season, August.

Remarks.—It bears abundantly. Nearly worthless.

Reine des Poires. Not suited to our soil and climate. Worthless here.

RONDELET. Color, yellow, with russet dots, in sun reddish; form, obovate; size, 3; use, table; texture, yellowish white, buttery, juicy, sweetish, and aromatic; quality, 2 to 1; season, September.

REMARKS.—Tree, a good bearer, upright, suited to orcharding. "Nearly best."—*From good judges at Cleveland*—among them Dr. Kirtland: A. H. Ernst does not like it. (Specimens of the same fruit will differ.)

ROPES. Color, russet and brown, a little tinge of red in sun; size, 3; use, table; texture, rather coarse, but melting; quality, 2 to 3; season, September.

REMARKS.—Name of a gentleman at Salem, Massachusetts. "Good."—*From authority of value.*

ROSS. Color, yellowish green and russet; form, obovate; size, 2; use, table; texture, rather gritty (this does not affect the *flavor*), sweet, juicy; quality, 2 to 3; season, Winter.

REMARKS.—Tree, of great vigor. Fruit, not remarkably good, but classed "good" by men of competent judgment.

ROUSSELET DE RHEIMS. See Spice, or Musk.

ROUSSELET HATIF. Color, green, with russet and red; use, table; texture, buttery, delicious; quality, 1; season, July.

REMARKS.—One of the best early pears. Early Catherine, of Philadelphia. Fine, rich, waxy, and luscious. Form, like a calabash, with a long curved neck, and a long fleshy stem. The skin is, on one side, of yellow, the other, a rich russet, or brownish red. The tree grows to a large size before it bears; it is then very fruitful. The limbs are long, and, when full of fruit, hang like a willow. This pear should be always suffered to hang on the tree till ripe. Coxe is, for this, our best authority, as for many older fruits of the Middle States, when we have not had opportunity of testing them with us, *on the spot.* This fruit was exhibited, however, by our indefatigable friend, A. H. Ernst, at our rooms, and at the Cincinnati Horticultural Exhibition, 1855. Fruit Committee consider it "an excellent pear."

Rousselet de Meister. Unworthy.

Rousselet Double Esperin. Color, dull greenish yellow, with cinnamon russet; form, obovate, acute pyriform; use, table; texture, melting, sprightly; quality, 2 to 3; season, September.

REMARKS.—"Very good."—*Wilder, in Horticulturist.*

ROSABIRNE. Color, russet, on dull greenish yellow; uneven surface; form, obovate, acute pyriform; use, table; texture, melting, juicy, brisk, sub-acid; quality, 2 to 1; season, September.

REMARKS.—"Very good," may be "best."—*Col. Wilder in Downing; and other Horticulturists.*

ROYALE D'HIVER, or *Winter Royal.* Color, yellow and red; form, pyriform; size, 1; use, table; texture, melting, sweet; quality, 2; season, November to January.

ROSTIEZER. Color, dull green; form, obtuse pyriform; size, 3; texture, juicy, melting; quality, 1; season, July or August.

REMARKS.—Fine fruit, rather small, delicious flavor; productive.

SALVIATE. Color, waxy yellow; form, nearly round; size, 2; use, table; texture, buttery; quality, 2; season, July or August.

SHELDON. Color, greenish yellow, with some russet; form, obovate, acute pyriform; size, 2; use, table; texture, melting, juicy, sugary, gritty at core; quality, 2; season, September.

REMARKS.—Foreign. Tree vigorous; tolerably productive.

SANS PEAU, or *Skinless*. Color, pale greenish yellow; form, oblate pyriform; size, 3; use, table; texture, white, juicy; quality, 2; season, July or August.

REMARKS.—A high flavored, early fruit. Exhibited by R. Buchanan, at the Cincinnati Horticultural Society Rooms.

SECKEL. Color, yellowish green, with red cheek, smooth; form, regular; size, 2; texture, very juicy, very sweet, and delightfully pleasant; quality, 1; season, August, September, and October.

REMARKS.—From Mr. Seckel, of Philadelphia. It is the finest pear of this or any other country. A native fruit, "well adapted for the vicinity of Cincinnati."—*F. G. Cary*. What more can be said of this superlatively choice fruit? "Best grafted on the apple (a singularity), reserving part of the apple branches to give large size and flavor." So says Mr. Gabriel Sleath, an experienced horticulturist and cultivator, and one of the pioneers of Cincinnati, on flowers and fruit, with Jas. Howarth, and, to be in good company, may I be permitted to add the author here. Tree very vigorous and beautiful on its own stock, and remarkably hardy, healthy, and luxuriant in growth, and requires on its own stock a liberal use of the knife. The pear tree is liable to be much injured, if pruned by those who do not understand the nature of it. The blossoms are commonly produced from buds at the extremity of the last year's shoots; and, as these are often cut off by the unskillful pruner, it prevents their producing fruit, and causes the boughs to send out new branches, which encumber the tree with too much wood. The Summer is the best time to look over pear trees, and to remove all superfluous and interfering shoots, which would too much shade the fruit, although in this climate they will stand more shade than in the north of Europe. If this be carefully done, they will require but little pruning in the Fall.—The original venerable Seckel pear tree stands in a meadow in Passyank township, less than a quarter of a mile from the Delaware River, opposite to League Island, not more than half a mile from the mouth of the Schuylkill, and about three and a half miles from the city of Philadelphia. It measures six feet in circumference one foot above the ground, and four feet nine inches higher up. It is about thirty feet high. It has its usual roundness. It is now in a great state of decay.

Seigneur d'Esperin. Color, lemon yellow; size, 2; use, table; texture, juicy, sweet, pleasant; quality, 1 to 2; season, August or September.

Remarks.—A new, beautiful, and very superior fruit. Tree, a healthy, upright grower.

Sinclair's Seedling.

Remarks.—A small, but excellent fruit, from Baltimore, Maryland; only rather liable to crack, as is complained of by some in particular situations, in the White Doyenne and a few others;—unsuitable soil and aspect probably the chief causes. Time and more experience will improve these things, by closer attention.

SKINLESS. See Sans Peau.

REMARKS.—A fruit of the size of the Early Catherine; skin smooth and very thin. Color, greenish yellow, with a little blush. The stem is long and small; the flesh juicy and breaking, rather than melting; of a pleasant, sweet taste; very attractive to wasps and bees, the thin skin much favoring their operations. The tree and foliage are of delicate growth. Season, about the end of July, or a little sooner sometimes, varying, of course, like other fruits, according to the seasons in our latitude, being sometimes a month or even six weeks, earlier or later. This has been borne in mind by the author in the statements of the seasons of all our fruits. We are also about a month earlier here than in the North or East; and still earlier than in the North of Europe. This changes their Winter fruits sometimes into Autumn with us, and their Fall fruits into nearly Summer, in the West; thus, often materially affecting their character and quality. The difference of our soil, too, having no slight influence on them.

SCHENCK'S, *or Hosenschenck* (Stocking Leg). Color, light yellowish green; form, obovate; size, 1; use, table; texture, tender and melting, with a juicy and very sprightly flavor; quality, 2 to 1; season, September.

REMARKS. — From Lancaster, Pennsylvania. Large. Yellow. Resembles Bartlett, and perhaps as good. Though this would be very doubtful, in the estimation of most.

SIEULLE. Color, pale yellow, with a little red on the sunny side; form, roundish flattened; size, 2; use, table; texture, buttery, melting, rich, and very good; quality, 2 to 1; season, September.

REMARKS.—A new fruit. From the London Horticultural Society's garden, which has performed immense benefit to pomologists in England, as well as to the world, although the results of their trials have but a partial bearing on the benefits to be derived from us here, on account of the great difference of climate, soil, and other circumstances. But the old mother country has undoubtedly ever taken the foremost lead in the improvement of the world in every thing. We are following, shall follow, and, in due time, under many more favorable circumstances, natural advantages in particular, will, in all likelihood, surpass Great Britain in almost every thing, pomology as well.

SOLDAT D'ESPEREN, *or Soldat Laboreur d' Esperen.* Color, greenish yellow, becoming yellow, with many patches, stripes and dots of russet; form, pyriform, sometimes ovate pyriform; size, 1; use, table; texture, coarse, juicy, melting, somewhat perfumed; quality, 1; season, November to January.

REMARKS.—Raised by Mr. Esperin, of Malines. Must not be confounded with Soldat Laboreur of the French, which is a third or fourth-rate fruit. Described by Col. Wilder in *Downing's Horticulturist.*

Soldat Laboreur of the French.

REMARKS.—Not worth notice. Unworthy.

Spanish Bon Chretien. Unworthy.

Spice, or Musk Pear. Color, greenish yellow; form, oval; size, 3; use, table; texture, breaking, or half buttery, with a sweet, rich, aromatic flavor; quality, 2; season, August.

REMARKS.—There is a pretty strong resemblance in the color, form, and flavor of the Seckel, but the latter is much the most delicious. Tree, very vigorous, grows with long shoots, like the Catherine. It does not bear till large; then very fruitful. Eaten when fully ripe, very good, but too small.

St. Ghislain, *or St. Galen.* Color, pale yellow; form, pyriform; size, 3; use, table; texture, juicy, sugary; quality, 2; season, August and September.

Remarks. — Tree, upright, vigorous, growth; young shoots, light brown. Requires a warm, rich soil; otherwise, a little insipid. Excellent, thrifty, and upright. Productive.

St. Andre. Color, yellowish green; form, globular, acute pyriform; size, 2 to 1; use, table; texture, juicy, tart; season, August and September.

Remarks.—Rather inclined to rot at core. Tree, vigorous, healthy grower. Early bearer.

ST. NICHOLAS. Color, greenish yellow; form, oblong pyriform; size, 1; use, table; texture, rich, sub-acid, slightly perfumed; quality, 2; season, Fall.

Remarks. — Flesh, with some astringency, next the skin. Described by Col Wilder, in *Downing's Horticulturist*, "As a standard good for market." This is an instance of a small pear in France, growing a large pear in this country.

ST. GERMAIN. Color, green; form, irregular; size, 1; use, table; texture, juicy; quality, 2.

Remarks.—Keeps pretty well with care. Stem short, and generally planted in an oblique direction. Flesh, highly flavored. Requires a warm situation to ripen.

ST. MICHAEL OF BOSTON, or *Virgalieu, Yellow Butter, or Beurre d' Oree.* Color, bright yellow; form, oblong;

size, 1; use, table; texture, juicy, melting, buttery; quality, 1; season, September to November.

REMARKS.—One of the very best. White Doyenne of France. Almost as good as the Seckel. Should be gathered before fully ripe, and kept some time in the house. It is an abundant bearer, and produces early.

STEVEN'S GENESSEE. Color, greenish yellow; form, round obovate; size, 1; use, table; texture, juicy, sugary, aromatic; quality, 2; season, August and September.

REMARKS.—Not equal to the above (St. Michael). Apt to rot at core. Native of Rochester, New York. Good on pear and quince, but the West does not seem to suit it very well, though it has done well in some parts of the country. Grows to a good size; of a beautiful gold color when ripe. It needs to be eaten as soon as ripe, as it soon gets mushy. The tree is a fine, upright grower; the fruit large and fine. The trees spread in their habits, and are only moderately productive.

Stone Pear. Color, greenish white, with large red blush; form, top-shaped, or turbinated; size, 1 (larger than Bartlett); use, baking; quality, 3; season, 10th September to 10th October.

REMARKS.—For market and baking only. Its fine appearance renders its merits deceptive. In passing this judgment on this fruit, that it is merely fit for sale, we advance a very poor encomium on the popular knowledge of pomology. Indeed it is rather low, and a very few years back was much lower. But ought not this fact to be a strong inducement to the philanthropist, and to horticulturists, to spur them on to still unceasing efforts for the improvement and establishment of fruits, not only in *quality* but in *quantity*, so that they be within the reach of all, at a reasonable price.

Styer. Color, green; form, round; size, 2; texture, buttery, juicy, gritty; quality, 1; season, August.

REMARKS.—Origin unknown.

Striped St. Germain.

REMARKS.—Only a striped variety—like the Striped Duchesse. Unworthy of much notice.

SUCRE VERTE, or *Green Sugar.* Color, green; form, oblong; size, 2; use, table; texture, buttery; quality, 2; season, September. Of vigorous growth.

Sugar Top. Unworthy.

SOUVRAIN D'ETE. Color, lemon yellow; form, round obovate; size, 2; texture, juicy; quality, 2; season, August.

REMARKS.—Described by *Col. Wilder, in Downing's Horticulturist.* Flesh, melting, tender, and very juicy; flavor, sprightly, a little vinous, rich.

SUMMER BERGAMOT. See Bergamot of Summer.

REMARKS.—Should be eaten before it is too ripe. The skin is green, full of small russet spots; but, when fully ripe, it becomes yellow. It is a highly flavored fruit, if gathered from the tree; but, when too ripe, it becomes dry, and loses its flavor. "Poor when too ripe." The size is small; of a round form; the flesh rich and sprightly. It is the least vigorous pear tree in our country, of moderate size, and great hardiness; free from blight. The fruit in perfection from the middle to the end of July."—*Coxe.*

Summer Franc Real. Color, dull green; form, obovate, obtuse pyriform; quality, 2; use, table; texture, buttery, juicy, sweet; season, August.

Remarks.—Foreign. Tree thrifty, hardy, good bearer; best on quince. "Indifferent in quality. May prove better another year."—*A. H. Ernst.* Probably a second-rate fruit. Sometimes a second-rate fruit is better than a first, it depending so much for what purposes they are required. If a person wants half a peck of the very best pears (speaking by way of comparison), for any particular purpose, or for his own use and enjoyment, then he may grow the best fruit; but if he wishes for half a bushel for market, to make money, it is another thing. In both cases the same amount of labor and pains are bestowed upon the culture.

Summer Pine Apple. See Ananas d'Ete.

Summer Rose. Unworthy.

Summer Colmar. Color, green, white at maturity; form, round, obtuse conical; size, 2; use, table; texture, melting, sugary, vinous; quality, 1; season, August.

Remarks.—An unsightly, but exceedingly luscious, and fine fruit. Tree, unsightly, rough, and brittle.

Summer Bon Chretien. Color, yellowish green; form, very irregular, oblong; size, 1; use, table; texture, melting, buttery; quality, 2; season, August and September.

Remarks.—Fine odor. Skin very smooth. It frequently cracks, which diminishes its excellence. Leaves large and smooth. Tree vigorous. Sometimes misnamed Jargonelle in this country. Exhibited by T. V. Petticolas. July 29, 1855. Fruit Committee report, "delicate and fine." Ripens about the time of the Seckel, and several other pears of high reputation. Mr. R. Buchanan ranks it as a good market fruit—not of the first order for market cultivation.

SUMMER DOYENNE. See Doyenne d'Ete.

REMARKS.—Exhibited by T. V. Petticolas, July 29, 1855. Fruit Committee considered it "beautiful and small—a good fruit."

SUMMER ST. GERMAIN. Color, pale green all over the surface; form, obovate; use, table; size, 2; texture, juicy, tender, sweet, with a very little acid, and very good; quality, 2.

REMARKS.—A pleasant, juicy, Summer pear, bearing large crops, and growing vigorously.

SUMMER PORTUGAL. Color, yellow; form, round, obovate; size, 3; use, table; texture, white, juicy, buttery; quality, 1; season, July and August.

SUPERFONDANTE. Color, pale yellow, marked with a few dots, and sometimes russet; form, obovate; size, 2; use, table; texture, buttery, melting, and good; quality, 2; season, September.

REMARKS.—Of the same class as the White Doyenne, and like it somewhat in appearance and flavor.

SULLIVAN. Color, pale greenish yellow; form, oblong pyriform; size, 2; use, table; texture, juicy, melting, sugary, and agreeable; quality, 2; season, August and September.

STYRIAN. Color, deep yellow, with a bright red cheek, and streaks of light russet; size, 2 to 1; use, table; texture, crisp, with a rich, high flavored juice; season, Sept.

SURPASSE VERGALIEU. Color, yellow, small dots; form, varying; size, 2; use, table; texture, buttery, juicy, and aromatic; quality, 2; season, September.

REMARKS.—Probably an American seedling. A regular abundant bearer, on pear or quince. A vigorous, healthy tree, with yellowish brown wood; growth upright. (This upright growth, of many kinds of trees, is advantageous for the smaller room they take up in gardens and grounds, while forming a pleasing variety of character to the scene.)

Swan's Egg. Color, green, part brown; form, elliptical; size, 2; use, table; texture, melting, musky; quality, 3 to 4; season, September.

REMARKS.—A good little pear in England, being very hardy, and bearing large crops as a standard, without needing training, which is most practiced there from the coolness of the climate in Summer, compared with ours in the Middle States. It is very little esteemed here, where they can have such fruits as the Seckel, Kirtland's, etc.

SWAN'S ORANGE, *or Onondaga.* Color, pale greenish yellow, becoming golden yellow when matured, a good number of russet dots, and at times a dull blush in the sun; form, ovate, obovate; size, 1; use, table; texture, white, juicy, buttery, and rich.

REMARKS.—Succeeds equally well on the pear or quince stock, and, as a profitable market variety, deserves extensive culture. It will compare well with the great Bartlett, or Williams' Bon Chretien (before described). It is not so sweet as many fine pears, but of a very good flavor, and large size. "It is very valuable."—*Dr. Warder.*

Sylvange is unworthy.

TRIOMPHE DE JORDOIGNE. Form, obovate; size, extra 1; use, table; texture, melting, sub-acid; quality, 1; season, September and October. Described by Col. Wilder in *Horticulturist*, by Downing. Very large, and promises well.

Tea. Color, rich yellow; form, round oval; size, 3; use, table; texture, white, melting, juicy; season, August.

REMARKS.—Origin, New Haven, Conn.

THOMPSON. Color, greenish yellow; form, turbinate, round to eye; size, 2 to 1; use, table; texture, buttery, melting, sugary; quality, 2; season, September and Oct.

REMARKS. — Flesh, white and greenish yellow, fine, melting, buttery, juicy, sweet, and highly and agreeably aromatic.

TYSON. Color, dull yellow and russet; form, round, pyriform, irregular; size, 2; use, table; texture, juicy, sugary, melting; quality, 1; season, July and August.

REMARKS.—From Pennsylvania. Tree of vigorous, upright growth, with reddish brown wood. This, with the Bartlett, Seckel, White Doyenne, Bloodgood, Louise Bonne de Jersey, Glout Morceau, Duchesse d'Angouleme, Julienne, Doyenne Robin, and some few others, Mr. Buchanan finds to succeed admirably with him, as dwarfs. He plants them round the borders of his garden, about eight feet apart. Thus they occupy but little room, and require but slight attention. Sometimes a sucker from the stock has to be cut away, or a straggling branch shortened in—and but little more. Too deep, or highly manured ground, makes them grow too much to wood, but *no fruit.* In this way, even in rather bad years, some of the trees will bear fifteen to twenty pears, when only five to six feet high. Thus they look promising, to a considerable extent, for amateur culture in our locality. The soil and climate seem to suit them tolerably well. For a *profitable* market crop, we shall probably have to rely principally upon our standard pear trees; but time will show—"we shall see." One thing may be here observed, viz: that owing to our dry Summers, compared with Europe and the Atlantic States, the fibrous small roots of the quince, which are near the surface of the ground, are more affected by drouth than the pear roots, which seek their sustenance deeper in the earth. This is rather against the dwarf pears for market use. If the pear is permitted to take root, then it will soon cease to be dwarf, and grow away high. With regard to the grafting of pear trees, on their own stock, and on the quince, it must not be forgotten that, in the former case, the union is more perfect than in the latter, where the sap is not in action at the same instant of time, and the quality of their secretions, be they what they may, can not be as perfectly identical, or precisely the same. This will, and does have a great bearing on the success of the culture of dwarf pears, compared with the standards; and on this account, also, we can not look for the same permanent prosperity of these trees. We must necessarily incur the loss of time and money in their continual decay and renewal. It is only when varieties of the *same species* are worked on each other, that a perfectly sound and durable union is effected, and not always, even then, as we see when a fast growing apple-tree is grafted upon a diminutive and slow-growing variety, such as the Paradise. A peach-tree, when budded on the varieties of the plum, were it not that the plum stock is impervious to the peach worm, would not be more durable, in this country, than when budded on its own stock. One very great advantage, which must not be overlooked, in the use of dwarf pears is, that they fill that vacuum in the cultivation of the pear which is so trying to the patience of fruit-growers, and which must necessarily exist, until the standards shall have come into full bearing.

UPPER CRUST. Size, 3; use, table; texture, buttery, melting; quality, 1; season, July.

REMARKS.—A South Carolinian. Tree, healthy, moderate grower. Fruit, size of Dearborn's Seedling.

URBANISTE, or *Beurre Picquery, etc.* Color, pale yellow, with gray dots, and some russet lines; form, obovate pyriform; size, 2; use, table; texture, white, yellowish at core, buttery, melting, vinous; quality, 2; season, September and October.

REMARKS.—Good. Tree, moderately vigorous, and hardy; well suited to Western soils. Not an early bearer, but when it does, produces regularly and abundantly.

Uvedale St. Germain. Poor. For baking only.

VALLEE FRANCHE. Form, round, obtuse, turbinate; size, 2; use, table; texture, poor; quality, 3; season, July or August.

REMARKS.—Exhibited to Cincinnati Horticultural Society, by A. H. Ernst, August 18, 1855. Committee on Fruit considered it "quite a second, or even third rate Summer pear." "Deficient in flavor and character."—*A. H. Ernst.*

VAN ASSCHE. Color, light yellow; form, obovate obtuse; size, 1; use, table; texture, buttery, melting; quality, 1; season, August and September.

REMARKS.—New pear. Worthy of general culture.

VAN ASSENE. Color, greenish yellow; form, round, flat at crown; size, 1 to 2; use, table; texture, juicy, melting, sweet; quality, 2; season, August.

REMARKS.—A new, and fine pear.

Van Buren. Poor. Unworthy.

VAN MONS LEON LE CLERC. Color, pale yellow, gold at maturity; form, ovate, obovate pyriform; size, 1; use, table; texture, buttery, melting, juicy; season, October to January.

REMARKS.—Foreign. Valuable on the quince. Bears early and well. Very large and fine, but the tree is always old. Has no stamina or vigor.

VERTE LONGUE. Color, green, even at maturity, with numerous minute dots; form, long pyriform, narrowing a good deal from the middle toward both ends; size, 2; use, table; texture, very juicy, with a sweet, slightly perfumed, very excellent flavor; quality, 2; season, August and September.

REMARKS.—Very pleasant and lively tasted fruit. Bears abundant crops. The true sort is large, green and good. We doubt if all our nurserymen have the real Mouthwater, one of its synonymes.

VERTE LONGUE, or *Panachee.* Color, yellow and green striped; form, round at blossom end; size, 2; use, table; texture, buttery; quality, 2; season, September.

REMARKS.—Striped sort. Bears abundantly.

VICAR OF WINKFIELD, *Le Cure, Clion, etc.* Color, skin fair and smooth, pale yellow, sometimes, with a brownish cheek, and marked with small brown dots; form, long, pyriform, often six inches, and a little one-sided; size, 1; use, table, but chiefly cooking; texture, very juicy generally, but it varies; sometimes it is buttery, often crisp, with a good sprightly flavor; quality, 1; season, October to January.

REMARKS.—The opinions differ more on this fruit than any other, unless it be Passe Colmar. A great many highly approve of it. It depends, we think, a great deal upon the way and time it is ripened. It certainly is the very best kind of baking pear; and, sometimes, we find it a fine table pear. It probably requires a warmer temperature to ripen than most other pears. It can not be easily bruised when picked, as it is very firm, but all pears can not be too tenderly handled and deposited. It is very productive, hardy and large. The fruit branches droop.

VIRGALIEU, see *White Doyenne, or St. Michael's.*

VIRGALOUSE. Color, yellowish, green at maturity, with gray or reddish dots; form, conical; use, table; texture, juicy and rich; quality, 1; season, October to January.

REMARKS.—It has brought twenty dollars per barrel at New York. Known as Butter, or St. Michael, at Boston. It is of good size, high flavored and juicy, and has every fine quality necessary. It should be picked some time before it is ripe, as it is a long while ripening. Exhibited by some member at the Horticultural Society Rooms. "A good fruit."—*Fruit Committee.*

WASHINGTON. Color, lemon yellow, with red russet; form, roundish ovate, or ovate pyriform; size, 2; use, table; texture, juicy, sugary; quality, 1; season, August and September.

REMARKS.—Washington deserves its name; it is excellent, thrifty, hardy and productive. American — from Delaware. Tree healthy, of moderate growth; an annual bearer. A young but moderate bearer. Fine for the amateur, particularly further South.

WENDELL. Form, round, obtuse; size, 2; use, table; season, August.

REMARKS.—"Flesh, white, breaking, tender, juicy and a little gritty, sweet, pleasant; somewhat aromatic."—*A. H. Ernst.*

WHITE DOYENNE, *Virgalieu, or Yellow Butter.* See *St. Michael's.*

REMARKS.—"Excellently well adapted to the locality of Cincinnati."—*F. G. Cary.* Should be picked before ripe. Does well on quince or standard. Sometimes it has cracked, but not lately; probably from atmospheric causes. Great sales of these trees in the nurseries. It is particularly well adapted for the quince stock.

WIEDOW PEAR. Color, yellowish green; form, regular turbinate; size, 2; use, baking; quality, 1; season, September and October.

REMARKS.—Skin thin; flesh, white, very melting, buttery, juice abundant, slightly acidulous, or vinous, agreeable, perfumed, and highly flavored. Described by Andrew Leroy in *Horticulturist.* A delicious, first-rate fruit.

Windsor, or Summer Bell.

REMARKS.—Good for nothing. Exhibited by M. McWilliams, July 29, 1855. Fruit Committee considered it "A poor fruit."

WILLIAMS' EARLY (not Bartlett, or Bon Chretien). Color, bright yellow, with rich scarlet dots on sunny side; form, roundish, turbinate, regularly formed; size, 3; use, table; texture, very juicy, half buttery, rich, with a slight musky flavor. (*Downing's Fruits and Fruit Trees of America*).

REMARKS. — Tree, a moderate grower.

WILBUR. Color, yellowish green; form, oval, obovate;

size, 2; use, table; flesh, white, juicy; quality, 2; season, August.

REMARKS.—From Somerset, Mass. Tree, a moderate grower.

WILLIAMS' VIRGALOUSE. Color, light green; form, irregular; size, 1; use, table; texture, juicy; quality, 2; season, November to January.

Wilkinson. Color, golden yellow; form, obovate, narrow near stalk; size, 1; use, table; texture, sugary.

REMARKS.—Rich yellow color. Flesh, yellowish white, fine grained, melting; and though rather solid in texture, juice sugary and vinous. Described by H. Wood in *Downing's Horticulturist.* A natural seedling. Sold at high prices in Fulton market, N. Y.

WINTER ROUSSELET. Color, lively russet; size, 2; use, table and baking; texture, sugary; season, long Winter keeper.

REMARKS.—An abundant bearer, and rather too acid for the table.

WINTER NELIS. Color, yellowish green; form, roundish obovate; size, 2; use, table; texture, juicy, sugary; quality, 1; season, October to December.

REMARKS.—Foreign. Very hardy. Prolific on pear or quince. Excellent, though Autumnal. Wood slender.

The following important conditions should be observed, as recommended by John Sayers, one of our most intelligent and reliable nurserymen, and florists, near Cincinnati, in the cultivation of the pear on the quince:

"1st. A proper selection of varieties, such as are known to succeed on the quince.

"2nd. Good, healthy trees worked on a good stock—the Angers, or Paris.

"3rd. A good, loamy soil, of moderate fertility, on a clay sub-soil, and from eighteen to twenty inches deep.

"4th. Planted so deep that all the quince stock is below the surface.

"5th. Ordinary good cultivation, and moderate pruning, but not pruned so severely as to deprive the plant of power to send down woody matter enough to keep the roots healthy and active. Nearly all the failures can be traced to a want of one or more of the above requisites for successful cultivation."

The truth is, our climate and soil is not so much against the success of pear, either on its own stock or the quince, as the want of attention to those particular requirements, which are necessary for each, in common with all other fruits, more or less.

WINTER ORANGE. Form, nearly round; size, 2; use, table; texture, melting, juicy; quality, 1; season, October to December.

REMARKS.—A very good Winter pear, when, in the West, we have but few to keep well.

WILHELMINE. Color, greenish yellow; form, round obovate; size, 2; use, table; texture, buttery, sugary, juicy; quality, 2; season, Winter.

REMARKS.—Foreign.

Winter Bergamotte. Color, russet; form, round, flat at ends; size, 2; texture, spongy; quality, 3.

REMARKS.—From England. Not of much value. "Good when taken just in time."—*A. H. Ernst.* To how many fruits does not this observation apply? There is one certain point in their perfection.

Zoar Beauty, or *Superb*. Color, light yellow; form varies; size, 2; use, table; texture, juicy, sugary; season, July and August.

Remarks.—Native of Ohio. Tree a vigorous grower, with dark brown shoots. This pear, from its great natural vigor, with some others like it, as the Seckel, Buffum, etc., would not require so rich a soil as pear trees do in general. The pear tree, as a general rule, requires a rather moist and tenacious soil; not, however, wet and saturated with stagnant waters. If placed on a loamy or clayey soil, abounding in the requisite inorganic elements (phosphate of lime and potash), with pure water percolating beneath at a depth at which it can merely be reached by the extreme roots, this tree will be as hardy, strong growing, and durable as the oak. The deficiencies which occur in most soils may be, to some extent, artificially supplied. Animal bones, urine, the sweepings and droppings of the roosting poultry, or of the poultry-house and yard, and guano, are the principal sources from which the surplus must be supplied.—Probably the very best mode of preventing *any species of blight*, either the *frozen sap blight*, the *canker blight*, often occurring in the *insect blight*, or the *fire blight*, is to bury about their roots large quantities of unground bones; time and weather breaking them down as rapidly as the trees call for supplies; the surface of the ground being also dressed with ashes and refuse lime. No one should set out one pear tree more than he can at suitable intervals cultivate with care, and can at certain times supply, in some form, with the requisite food.

FRUITS OF OHIO.

Statement of R. BUCHANAN, A. H. ERNST, *and* J. A. WARDER, *of Cincinnati, Hamilton county, Ohio, as reported to the American Pomological Society, at their annual meeting, held at the City of Boston, in September*, 1854.

PEARS.

"Some varieties of this fruit, as the 'Bartlett' and 'Seckel,' bear as well as the apple, and others one year in two or three.

"The Committee is largely indebted to one of its members, Mr. Ernst, for valuable notes on this fruit, carefully prepared, from his own experience in its culture.

"Many varieties, particularly those of American origin, thrive well as standards; but, as a general rule, the foreign sorts do best on the quince stock. The cultivation is principally in the hands of amateurs, but the high prices obtained for the pear in our markets will soon cause a more general culture, which is invited by our favorable soil and climate.

"The following are considered best: Bartlett, Beurre d'Aremberg, Beurre Benoist, Beurre Diel, Beurre Spence, Bloodgood, Dearborn's Seedling, Dix, Doyenne d'Ete, Duchesse d'Angouleme (very early), Easter Buerre, Flemish Beauty, Heathcot, Julienne, Lawrence, Louise Bonne de Jersey, Madeleine, Onondaga, Osborn, Pratt, Saint Ghislain, Seckel, Stevens' Genessee, Stone, Tyson, Van Assene, Washington, White Doyenne, Zoar.

"The following are rejected, as unsuited to this region, or for inferior size and quality: Amire Johannet, Beurre d'Amanlis, Beurre Capiaumont, Chelmsford, Colmar Neil, Early Catherine, Grosse Calebasse, Jargonelle, Moon's Pound, Musk Summer Bon Chretien, Orange Bergamot, Petit Muscat, Red Bergamot, Rondelet, Summer Franc Real, Valle Franche, Windsor."

INDIANA PEARS.—Recommended by Henry Ward Beecher.

Summer, or *such as ripen from the first of July to the last of August.*—Madeleine, or Citron des Carmes, Bloodgood, Summer Franc Real, Dearborn's Seedling, Julienne, Williams' Bon Chretien.

Autumn, or *such as ripen from September to the last of November.*—Stevens' Genessee, Belle Lucrative, Henry the Fourth, Washington, Dunmore, St. Ghislain, Seckel, Beurre Bosc, Andrews, Marie Louise, Doyenne or Fall Butter, Dix, Petre, Duchesse d'Angouleme.

Winter, or *those which ripen during the Winter and Spring months.*—Beurre Diel, Hacon's Incomparable, Passe Colmar, Beurre Ranz, Columbia, Beurre d'Aremberg, Van Mons' Leon le Clerc, Beurre Easter, Chaumontelle, Glout Morceau, Prince's St. Germain, Winter Nelis.

Those who wish only *four* trees, may select Bloodgood, Williams' Bon Chretien, Duchesse d'Angouleme, Beurre d'Aremberg. Those who have room for *eight*, to the above may add Beurre Bosc, Passe Colmar, Columbia, Winter Nelis. Those who wish *sixteen* trees, to the above may add Madeleine or Citron des Carmes, Summer Franc Real, Dunmore, Seckel, Dix, Beurre Diel, Beurre Ranz, Beurre Easter.

List of Pears for Southern and Central Ohio, Indiana, Illinois, Southern Iowa, Northern Kentucky, Maryland, Tennessee, and Northern Missouri.

Ananas d'Ete. Stock, pear; use, table and market; season, August.

Bartlett. Stock, pear; use, table and market; season, August.

Belle Lucrative. Stock, pear and quince; use, table and market; season, September.

Beurre Langlier. Stock, pear and quince; use, table and market; season, October and November.

Beurre Moir. Stock, quince; use, table and market; season, October.

Beurre Bosc. Stock, pear; use, table; season, September and October.

Beurre d'Anjou. Stock, pear and quince; use, table and market; season, October.

Beurre Easter. Stock, quince; use, table and market; season, January to March.

Buffum. Stock, pear; use, table and market; season, September.

Black Worcester. Stock, pear; use, cooking; season, November to January.

Bloodgood. Stock, pear; use, table and market; season, July.

Bon Chretien Fondante. Stock, quince; use, table and market; season, September.

Dearborn Seedling. Stock, pear; use, table; season, July.

Doyenne, White. Stock, pear and quince; use, table and market; season, September and October.

Fulton. Stock, pear; use, table and market; season, October.

Flemish Beauty. Stock, pear and quince; use, table and market; season, September.

Glout Morceau. Stock, quince; use, table and market; season, November to January.

Jalousie de Fontenay Vendee. Stock, quince; use, table and market; season, September.

Louise Bonne de Jersey. Stock, quince; use, table and market; season, September.

Lawrence. Stock, pear; use, table and market; season, November.

Long Green, of Coxe. Stock, pear; use, table and market; season, September and October.

Lewis. Stock, pear; use, table and market; season, October to January.

Nouveau Poiteau. Stock, pear and quince; use, table and market; season, October and November.

Doyenne d'Ete. Stock, pear; use, table; season, July.

Beurre St. Nicholas. Stock, quince; use, table and market; season, October and November.

Pound. Stock, pear; use, cooking; season, November and December.

Soldat Laboureur d'Esperen. Stock, quince; use, table and market; season, November and December.

Urbaniste. Stock, quince; use, table and market; season, September and October.

Vicar of Winkfield. Stock, quince; use, market and cooking; season, October to December.

Seckel. Stock, pear; use, table and market; season, September and October.

Stevens' Genessee. Stock, pear and quince; use, table and market; season, September.

Tyson. Stock, pear and quince; use, table and market; season, August.

St. Ghislain. Stock, pear and quince; use, table and market; season, September.

Winter Nelis. Stock, pear; use, table and market; season, November to January.

Jaminette. Stock, pear; use, table and market; season, November to February.

Payency. Stock, quince; use, table and market; season, September.

Washington. Stock, pear; use, table and market; season, August and September.

ZOAR BEAUTY. Stock, pear; use, table and market; season, July and August.

BEURRE D'AREMBERG. Stock, pear; use, table and market; season, November to January.

SKINLESS. Stock, pear; use, table; season, July and August.

ROSTIEZER. Stock, pear; use, table and market; season, August.

Imperial Society of Horticulture of the Rhone (France), upon Pears.

We most sincerely hope that Pomological Societies may continue to flourish and to labor unceasingly. As, sometimes, so much perseverance is required to completely settle the true merits of any sort of fruit, the work of these Societies will never cease, and but little diminish in their hands. There will be a necessity often for some reconsiderations. The very same specimen will vary from tolerably good to very good, and from good to indifferently good, and also from being grafted on a different stock, etc., etc. The different seasons will also occasion them to vary much; and sometimes, if many sorts are grafted on one tree, those which have less vigor than the others will become greatly deteriorated. The great varieties of soil, climate, and exposure, will change considerably the same kinds, so much so that the same fruit in one locality can be hardly recognized in another. We make these remarks in order to direct the attention of our readers to the fact that the Imperial Society of Horticulture of the Rhone (France), have communicated to the various Horticultural Societies of the world that the French Pomological Congress, touching the action of that body on the nomenclature of fruits, propose to abolish synonymous names of fruits, and give each fruit a single definite denomination, or title. Our Horticultural Society at

Cincinnati have resolved to coöperate with them as much as possible in this respect, in a matter in which they and the other Societies of America have been laboring for some years. The first session was held at Lyons in September last (1856), and it may be interesting, here, to enumerate the number and kinds of pears admitted into cultivation. There were sixty-three for standards and dwarfs, seven especially for espaliers (a common mode for gardens in Europe), eight varieties for the kitchen, and twelve especially for standard, or orcharding. The names of the first sixty-three are: Adele de St. Denis (synonymes follow, which are abolished), Alexandrine Droullard, Arbre Courbe, Beau present d'Artois, Bergamotte d'Angleterre (Gansel's Bergamotte, etc.), Bergamotte d'Esperen, Beurre Beaumont, Beurre Benoit, Beurre Bretonneau, Beurre Capiaumont, Beurre Clairgeau, Beurre d'Amanlis panachee, Beurre d'Aremberg, Beurre d'Anjou, Beurre Davy, Beurre de Nantes, Beurre d'Hardenpont (Glout Morceau, etc.), Beurre Diel, Beurre Giffart, Beurre Picquery (Urbaniste, etc.), Quetelet, Beurre Six, Bezy de Montigny, (not the Musk Doyenne, commonly so called), Bon Chretien Napoleon (Charles of Austria, etc.), Bon Chretien William (Bartlett, etc.), Bonne d'Ezee, Calebasse Bosc (Thompson), Calebasse monstre (Van Marum, etc.), Colmar d'Aremberg, Conseiller de la Cour, Cumberland, Des Deux-Sœurs, Delices d'Hardenpont d'Angers, Delices de Lowenjoul, Doyenne Boussoch, Doyenne Defais, Doyenne d'Hiver (Bergamotte de la Pentecote), Duchesse d'Angouleme, Duchesse panachee, Duchesse de Berry d'Ete, Epine du Mas, Esperine, Figue (Figue d'Alencon), Fondante de Charneux, Fondante de Noel, Belle ou Bonne, etc., Grand Soleil, Graslin, Jalousie de Fontenay (Jalousie de Fontenay Vendee, etc.), Louise Bonne d'Avranches (Louise de Jersey), Marie Louise Delcourt (Marie Louise Nova), Nouveau Poiteau, Passe Colmar, Rousselet d'Aout, St.

Michel-Archange, St. Nicholas, Seigneur (Esperen), Fondante d'Automne, Belle Lucrative, etc.), Shobdencourt, Soldat Labourer, Suzette de Bavay, Triomphe de Jodoigne, Van Mons (Van Mons de Leon Leclerc). The Congress observe of the sub-varieties; Panachees, Beurre d'Amanlis and Duchesse d'Angouleme, are rather small of their kind. Pears for espaliers are seven: Bergamotte Crassanne (Cressanne, Cresane d'Automne, etc.), Beurre Gris, Bezy de Chaumontel (Chaumontel, vulgarly Charmontel), Bon Chretien de Rans (Hardenpont of Spring, etc.), Doyenne Blanc, or White Doyenne (St. Michel, Beurre Blanc, etc.), Doyenne Gris, St. Germain d'Hiver. The varieties, the fruits of which are for cooking, are eight, viz., Belle Angevine (Angora, Bolivar, etc.), Bon Chretien, d'Hiver, Catillac (Grand Monarque, Teton de Venus, etc.), Certeaux d'Automne, Monsieur le Cure, (Le Clio, Vicaire de Wakefield, (query, Winkfield, etc.), Leon Leclerc, Martin Sec (Rousselet d'Hiver), Messire Jean. Pears specially for standards (orchard trees), Bergamotte Sylvange, Beurre d'Angleterre (Poire Anglaise, St. Francois, etc.) Beurre Goubault, Beurre Millet, Blanquet, Citron des Carmes (Petite Madeline, St. Jean, etc.), Doyenne de Juillet, Epargne (grosse Madeleine), Josephine de Malines, Rousselet de Rheims (Petit Rousselet, Rousselet musque), Seckle (Shakespear, Seckle pear), Zephirin Gregoire.

In closing our remarks on pears, we will observe, here, that one of the reasons why there has been a discouraging impression made as to the value and success of the Dwarf pear in the West and in this vicinity, is that some of the Eastern nurserymen, before they became acquainted with the right kind of stock, sent trees worked on the Upright Quince, instead of the Angers or Paris—the latter is the stronger of the two, if there is any difference.

PEACHES.

ALBERGE. Form, roundish; leaves, globose; flowers, small; flesh, yellow, red at stone; color, purplish cheek; size, 2; quality, 2; season, August; freestone.

REMARKS.—Foreign. Only second-rate in flavor.

ALLEN. Form, roundish leaves, globose; flowers, small; flesh, pleasant vinous flavor, juicy; color, white, red cheek; size, 3; quality, 2; season, September, freestone.

REMARKS.—Hardy, and good bearer, and eminently productive. Has been raised forty years from the seed, uniformly true. Cultivated by several of its name, in Walpole, Mass. (Cole). This peach, like the Blood, and many others, reproduces the same from the seed (there are also some plums which do the same). Probably the trees are more healthy and hardy in consequence.

Anne, Early. Form, roundish; leaves, serrated, without glands; flowers, large; color, white, with a faint, but beautiful tinge of red, next the sun; size, 2 to 3; quality, 2 to 3; season, 1st July; freestone.

REMARKS.—An old English sort. It is the first peach, of any value that ripens; except the Early Tillotson, which ripens at the same time, and is superior. The Early Tillotson, will, therefore, take its place for general culture. The Early Red and White Nutmeg, are not worth much, being too small. The trees of the Early Anne are of rather feeble growth.

BALTIMORE BEAUTY, *or Belle.* Form, roundish

oval; leaves, with globose glands; flowers, large; flesh, yellow, but red at the stone, sweet, and very good—a little mealy, if over-ripe; color, deep orange, with a rich, brilliant red cheek; size, 2 to 3; quality, 2; season, July; freestone.

REMARKS.—A very good, and remarkably handsome, peach, of native origin. Its chief fault is not being quite juicy enough.

BARRINGTON. Form, roundish ovate, apex rather pointed, suture on one side, moderate; leaves, crenate, with globose glands; flowers, large; flesh, slightly red at the stone, juicy, rich, and of high quality; color, nearly white, with a deep red marbled cheek; size, 1; quality, 2; season, July and August, freestone.

REMARKS.—Is best south of New York city. The fruit ripens at the medium season—a week after Royal George.

BELLEGARDE, *Violette Hative, French Royal George, Large Violette, Brentford Mignonne, etc.* Form, round and regular, suture, shallow, with a projecting point; leaves, with globose glands; flowers, small; flesh, a little red at stone, a little firm, but melting, juicy, rich, and high flavored; size, 1; quality, 2; season, August; freestone.

REMARKS.—One of the best that supplies the Paris market; also prized by the English.

BEER'S LATE RARERIPE.

REMARKS.—Succeeds well at Frankfort, Ky.

BELLE DE VITRY, synonyme, *Admirable Tardive.* Form, approaching oblate, apex depressed, suture deep; leaves, serrated without glands; flowers, small; flesh, rather firm, red at the stone, juicy, and rich; color, nearly white, tinged and marbled with bright and dull red; size, 2; quality, 1; season, September; freestone.

Remarks.—This is distinct from the Late Admirable and Early Admirable, often called the same, but which is six weeks earlier. Both of the latter have crenate leaves, with globose glands. This is a great fruit at Frankfort, Ky. Flesh, firm.

BERGEN, *or Bergen's Yellow.* Form, globular; leaves, or glands, reniform; flowers, small; flesh, yellow, melting, luscious; color, deep orange, with a broad red cheek; size, 1; quality, 1; season, August; freestone. It differs from the Yellow Rareripe; which, however, it much resembles in its more oblate form, deeper color, better flavor, and ripening about ten days later, and in its glands being uniform. A moderate and regular bearer. It is probably the finest of all the peaches, with yellow flesh. It came from Long Island, N. Y.

Bledsoe's Seedling. Form, roundish oblong; leaves, glandless; flowers, large; flesh, mild, sweet, grateful; color, red and yellow; size, 2; quality, 2; season, September.

Remarks.—One of the best peaches in Kentucky—at Frankfort.

Blood Clingstone. Form, suture, distinct; leaves, glandless; flowers, large; flesh, deep red throughout, firm, juicy; color, dark, clouded purplish red; size, 1 (sometimes twelve inches round); quality, 1 (for pickling and preserving); season, September; clingstone, as its name denotes.

Remarks.—For pickling and preserving only. The old French Clingstone is not so large. There is a Blood Freestone, a variety of this, size 2, leaves without glands, and not worth as much.

BRECKENRIDGE, see *Old Mixon Cling*.

REMARKS.—A great peach at Frankfort, Kentucky, and a valuable variety everywhere, as well as the Old Mixon Free; both good kinds for market, especially the latter.

BREVOORT. Form, round and slightly oblate, suture distinct, deep at apex; leaves, with reniform glands; flowers, small; flesh, rather firm and slightly red at stone, rich, sweet and high flavored; color, nearly white, with a faint cloudy tinge, a bright red cheek; size, 2 to 1; quality, 1; season, August or September; freestone.

REMARKS.—Very good for the garden. From New York. A choice variety.

CABLE'S LATE MALACATUNE. Freestone; yellow.

REMARKS.—Is ripe about a week after Crawford's Late. It is a seedling from the old Red Cheek Malacatune, or Yellow Malacatune, the parent of most of the new yellow peaches lately known.

CARPENTER'S RED RARERIPE. Form, roundish; leaves, reniform glands; flowers, small; flesh, melting, juicy; color, red and white; size, 2; quality, 2; season, September; freestone.

REMARKS.—A great peach at Frankfort, Kentucky.

CLARK'S EARLY.

REMARKS.—A small native red peach, of good appearance, and of lively and decided rich flavor; the earliest on the list; tree of rather slow growth; productive; fruit ripe about the 28th of July; originated in St. Louis, and named by the Chairman of the Society in honor of Mr. Lewis Clark, who raised it.—*Fruits of Missouri*, by Thomas Allen, of St Louis.

Cole's Early Red. Form, roundish; leaves, or glands, globose; flowers, small; flesh, good, but rather dry; color, pale yellow, mostly covered with red; size, 2; quality, 2 to 3; season, August; freestone.

REMARKS.—American. Productive; not much can be said in its favor, beyond this—at any rate, for the West, where we have so many finer seedlings; and to these the year 1855 witnessed a very valuable accession, a list of which is given at the end of the description of this fruit, which see.

COOLIDGE'S FAVORITE, or *Coolidge's Early Red Rareripe.* Form, round, largest on one side; leaves, or glands, globose; flowers, small; flesh, melting, rich, juicy; rich, sweet, delicious flavor; color, white, with a bright red blush, generally mottled; size, 1 to 2; quality, 1 to 2; season, July and August; freestone.

REMARKS. — Tree, stout, healthy and prolific bearer. Rather too tender for market. Much eaten by bees, wasps, etc., on account of its delicate texture; and, from the same cause, injured greatly by wet weather when ripening. We can recommend it to amateurs and for small gardens, yet there are others which can be better depended upon. From Watertown, Massachusetts. It is named after a gentleman residing there.

COLUMBIA. Form, round, with a shallow crease, or suture all round; glands, or leaves, globose; flowers, small; flesh, yellow, often shows a red streak next the skin; color, dull and dingy red, curiously marked and striped with dark red; size, 1; quality, 1; season, August; freestone.

REMARKS.—Origin, New Jersey. Shoots, dark reddish purple. It is a singular and peculiar fruit. Mr. Coxe, one of the best and earliest writers on fruit in this country, very practical, and whose work is now becoming scarce, raised it.—Coxe's work abounds with good shaded likenesses, engraved on wood, of all of the first-rate fruits of America. Mr. Coxe considers the texture of this peach very like a pine apple, rich, juicy, and of very excellent flavor. Very fine. Specimens of this fruit were exhibited at our Fall Exhibition, in 1855 and 1856, by Mr. Bush of Covington, and others. Mr. Bush has succeeded well with many fine fruits, and particularly with plums, last year (1856) saved from the Curculio in a yard where poultry roosted and ran. His Raule's Janets are always splendid, with several other kinds. This may, in part, be attributed very much to the favorable soil and hilly situation in Kentucky, adjacent to Cincinnati. It is greatly favorable, also, to the growth of strawberries and raspberries. The hills are very healthy, and what is healthy for animal life, has the same effect, in a great measure, upon vegetable tissues.

CRAWFORD'S EARLY. Form, obovate, ovate; leaves, globose, in their glands; flowers, small; flesh, yellow, juicy; color, yellow, red cheek; size, 1; quality, 1; season, August; freestone.

REMARKS. — Of American origin. For market it is planted largely by all cultivators as the best of the yellow fleshed varieties. It is a very prolific bearer. The fruit is rich and sweet, if it has the benefit and advantage of a sunny exposure; otherwise, in the shade, it is slightly acid. This is another of the most valuable kinds raised in this country, from seed. It is not so common here to raise this, or any other fruit, by the ingenious method of impregnating the blossoms. But, as time advances, there will, it is hoped, be more of this done with many fruits, as the grape, etc. There is a singular fact related in *Philips' Companion to the Orchard*, published in 1831, as follows: "T. A. Knight, President of the London Horticultural Society, procured a new peach by this operation: he impregnated the pistillum of the blossom on an almond tree, with the pollen of the peach flower; and this almond, when planted, produced a peach tree instead of one of its own kind, and has since ripened peaches.—The oldest trees in England, of from forty to sixty years, generally yield a good crop, when younger ones fail; the finest peaches having been gathered from trees of the greatest age. How different from our climate!

CRAWFORD'S LATE. Form, round; suture, shallow; leaves, crenated with globose glands; flowers, small; flesh, red at the stone, juicy, vinous, hardly first-rate; color, yellow, with a broad dark red cheek; size, 1; quality, 1 to 2; season, September and October; freestone.

REMARKS.—This deserves a place in all collections. Productive. It is among the first as a late variety for market. Origin, New Jersey. The Red Cheek Malacatune is mistaken for it in some localities. Suits the Northern, Middle and Western States, as a market variety, as well as Crawford's Early.

Early Anne. See Anne Early.

EARLY RED. See Cole's.

EARLY TILLOTSON. See Tillotson.

EARLY TROTH. See Troth's Early.

Early Admirable (not *Belle de Vitry*, nor *Admirable*). Form, nearly round; leaves, crenate, with globose glands; flowers, large; flesh, red at the stone, juicy, rich, sweet, fine; color, nearly white, with a red cheek; size, 2; quality, 2; season, June and July.

REMARKS.—Quite early, ripening very soon after Serrate Early York. French origin.

EARLY NEWINGTON; freestone. Form, roundish, with a distinct suture; leaves, with reniform glands; flowers, small; flesh, white, but red at the stone, to which many tentacles adhere, particularly if not fully ripe; color, pale yellowish white, dotted and streaked with red, with a rich red cheek; size, 2 to 1; quality, 1; season, after the Early York, about the latter end of July, or beginning of August. Supposed to be American.

EARLY NEWINGTON; clingstone; or *Smith's Newington*, of the English. Form, rather oval, narrower at the top, and one half a little enlarged; leaves, serrated, without glands; flowers, large; flesh, firm, pale yellow, but light red at the stone, firmly adhering; color, pale straw color, with a lively red cheek streaked with purple; size, 2; quality, 2; season, first of August.

REMARKS.—Not much cultivated in this country, where we have many better; yet it is one of the best early clingstone peaches. It is of English origin. With regard to the effects of soil, climate, and location, on fruits, nearly all the books which have been written in America, have taken by far too large a range of region, or country, embracing, indeed, the whole United States, and some of those which were but lately Territories. It is seldom that these works have been particular enough in their pages to designate any particular portion of the whole country, to which any fruit was specially adapted.

And yet any person, who has paid the slightest attention to the growing of fruit, must certainly be well aware, that from one hundred to two hundred miles, generally, forms the diameter of a circle in which many kinds of fruit can be profitably and successfully raised. It is really necessary that every locality should have its own fruits minutely and carefully described, and every thing relative to their character, quality, size, use, season, appearance, color, flavor, their popularity in market, keeping, etc., etc., most accurately pointed out. In the first place, the soil has a powerful influence in the modification of fruit—to such a degree that some are of fine size, and of the greatest excellence, in one soil, and of very little, if any value in another. It also changes the time of their maturity, and has a great effect on the vigor and health of a tree, its size, and age, or longevity. Climate, also, has a most important effect upon both trees and fruit. Some kinds will succeed only where they have been raised, not being able to endure, without much injury, their removal two hundred, or even one hundred miles, in any direction, from their native place; while others, again, seem to succeed well in almost every latitude, and even in foreign lands. Most of the apples of Europe are failures with us; and how much is it desirable that we should have a most critical acquaintance with them, so as to be able to point out to new-beginners the fruit which will answer their expectations, satisfy their palates, assist their pockets, and save their time.

Early Sweetwater. Form, roundish, with a slight suture; leaves, with globose glands; flowers, large; flesh, white, melting, juicy, sweet, and pleasant; color, pale white, very seldom with a faint blush when fully exposed; size, 2; quality, 2; season, July and August; freestone.

Remarks.—Early. Ripens not long after the Early

Anne, and two weeks earlier than the Early York. American, from the seed of the Early Anne, to which it has been compared, but it is larger and better.

EARLY YORK, or *Serrate Early York, or True Early York.* Form, round, ovate; leaves, serrated, without glands; flowers, large; flesh, greenish white, tender, melting, filled with rich, sprightly juice; color, pale red, dotted or greenish white in shade; size, 2; quality, 1; season, July; freestone.

REMARKS.—The Coolidge's Favorite, Royal George, and some others, come after this fine fruit. There has been a considerable mixing up, and difficulty with the proper nomenclature of this peach, and others, similar to it. It has been thought by some to be the New York Rareripe, and the Large Early York, both of which are distinct kinds. Tree, hardy and productive. It holds a high rank among American cultivators. It is the first really fine, early peach, which ripens.

GEORGE THE FOURTH; with twenty-five synonymes. Form, round; glands, globose; flowers, small, dull red; flesh, pale red at stone; color, yellowish white, with bright red dots; size, 2 to 1; quality, 1; season, July and August.

REMARKS.—Probably the greatest peach for amateur culture in the United States. It is of large size. Its flavor is very high. The tree is so healthy and so productive, that it is suited to all parts of the Union, as the National Pomological Society has decided. No garden is complete without it. It is an American seedling, from Mr. Gill, Broad street, New York.

GROSSE MIGNONNE. Form, round; glands, globose; flowers, large; flesh, yellowish white, marked with red at the stone, melting, juicy, with a very rich, high, vinous flavor; color, dull greenish yellow, mottled with red, and having a purplish red cheek; size, 1; quality, 1; season, beginning of August; freestone.

REMARKS.—The universal, high estimation in which this celebrated peach is held, may be known, if from nothing else, like the Bartlett pear, from the great number of different names, or synonymes which are attached to it. It is a good and regular bearer, a large and handsome fruit, and flourishes well, even in rather uncongenial climates, like Boston, and in nearly all soils.

HEATH CLING, *Late Heath, or Heath.* Form, oblate, narrow at both ends; glands, serrated; flowers, small; color, greenish white; size, 1; quality, 1; season, September and October.

REMARKS.—American. A very valuable kind in the West and South. Used most for preserves. Of a juicy, rich, luscious flavor. It often reproduces itself from the seed. Trees very hardy, and producing often when others fail. The fruit will keep long, in a cool room, wrapt in paper. This peach is at the greatest perfection in the State of Maryland. It is equal to the very best when perfectly ripe, and the best for preserving. The juice is most plentiful.

HINE SEEDLING. Form, round; leaves, globose; flowers, large; flesh, juicy, sweet, good; color, red and yellow; size, 1; quality, 1; season, September; clingstone.

REMARKS.—American; from the Heath Cling, but of finer color, and believed to be earlier. Raised by Daniel Hine, of Talmadge, Ohio. Fruit premium awarded to it by the Ohio Pomological Society.

HONEST JOHN, see *New York Rareripe, La Grange.*

Form, oblong, shaped somewhat like the Heath Cling; color, greenish white, generally some red on the side of the sun; leaves, with reniform glands; flowers, small; flesh, pale, melting, juicy, high flavored, delicious and sweet; size, 1; quality, 1; season, October; freestone.

REMARKS.—Part of its great value is its very late maturity. It has also fine flavor. It is valuable both for the table and preserving, the latter for its lateness.

HYSLOP. Form, oblate roundish; glands, reniform; flowers, small; flesh, juicy, melting, rich, luscious; color, white, deep rich red cheek; size, 1; quality, 1; season, September; clingstone.

REMARKS.—For Northern climates this is a most desirable fruit—more so than the Heath Cling. The latter is best for us, and nothing of the kind is capable of taking its place, from its peculiar qualities, lateness and firmness, juiciness, etc., for preserves in particular. Trees of the Hyslop prolific, healthy and most hardy.

INCOMPARABLE, *Pavie Admirable, Late Admirable, Cling.*

REMARKS.—This is only worthy of cultivation for market. Season, September.

JAQUES, *Jaques' Rareripe.* Form, round; glands, reniform; flowers, small; flesh, yellow, red at stone; color, downy dull yellow, with a red cheek; size, 1; quality, 1; season, August; freestone.

REMARKS.—A good market variety. Not of the highest flavor, but very saleable. A very sure bearer. Juicy, rich, slightly sub-acid. Vigorous and productive. There are various opinions regarding the character of this peach. It is good in some places and indifferent in others. It is not so easy to determine the characters of peaches and pears as apples—so many new are continually coming in.

Some that have come in with good names and a high reputation elsewhere, have been discarded. A few should be our main dependence. Every book on this subject strongly recommends some particular kinds. People expect that nurserymen should grow all those which are recommended. They go to work and get up all this stock for three or four years, and, by the time they are ready, something has been found wrong about them; others come into notice and are praised; these again fail in some way, and it is, therefore, a hard task for growers to keep up with the fashions in this way. The fact is, a few very good and well tried kinds should embrace our main supply and dependence.

LARGE EARLY YORK. See Early York.

Large Early. Not a synonyme of Large Early York. It is not so early as that fruit. It is of very rich, delicious flavor. Color, whitish, red cheek, purplish in the sun; stone very small; size, 1; quality, 2; season, August and September.

Large White Clingstone. From New York. Adapted to the Middle States. Excellent for preserves. Season, September.

LATE ADMIRABLE. Form, roundish, slightly oval; large suture, small point at top; glands, globose; flowers small; flesh, greenish white, red at the stone, melting, very juicy, and most delicious; color, yellowish green, pale red cheek, marbled with dark red; size, 1, very large; quality, 2; season, August and September.

Remarks.—Fine for a private garden; rather too delicate for carriage; but very deservedly popular among amateur fruitiers. Origin, France.

LATE RED RARERIPE, *Prince's Red Rareripe.*

REMARKS.—One of the best for general culture; season, August.

LEMON CLINGSTONE.

REMARKS.—Large and popular for market. Native of South Carolina. Season, September. Clingstone peaches are not generally appreciated. Certainly, in comparison with them for immediate, and convenient and delicious eating, the freestone peach, provided it is sufficiently juicy, is generally preferred. But for some culinary purposes, they are greatly superior to freestones. There is a firmness, substance, flavor, and juiciness possessed by them which gives them advantages when made into pies and pickles. They may be sent to more distant markets, or kept longer at home, than other varieties. They may be picked and remain several days before fully ripe, while freestones have to be hurried off for immediate sale. The Late Heath Clingstone can be picked a little time before frost, and kept sometimes several weeks on shelves in a cool place, and for a much longer time in a Schooley Fruit House—a considerable time into Winter.

MALACATUNE. Form, round, ovate; leaves, globose; flowers, small; flesh, deep yellow; color, yellow, deep red cheek; size, 1; quality, 1; season, August; freestone.

REMARKS.—American. Popular everywhere. A fine and lovely fruit. Some of the finest peaches have been derived from it—as Crawford's Early and Late, etc. This is of Spanish origin.

MORRIS RED RARERIPE. Form, round; glands, or leaves, globose; flowers, small; flesh, greenish white, red at stone; color, greenish white, red cheek; size, 1; quality, 1; season, July; freestone.

REMARKS.—Everywhere valued greatly; tree prolific, strong, and healthy. Origin, Philadelphia. Ripens a few days later than George IV.; bears heavier crops and fairer fruit.

MORRIS WHITE. Form, oval; glands reniform; flowers, small; flesh, white to stone; color, downy, greenish white; size, 1; quality, 2; season, August; freestone.

REMARKS.—Best for the South and South-west; more valuable North for preserving and cooking. Moderately prolific. Tree of great vigor and strength.

MONSTROUS CLING. Form, round obovate; glands, reniform; flowers, large; flesh, yellowish white, deep red at stone; color, yellowish white, some red; size, 1, very large; quality, 2; season, September and October; clingstone.

REMARKS.—Good for market profit, on account of its great size; requires a soil that is both deep and very fertile. Of foreign extraction.

MORRIS RED, *New York Rareripe, or Honest John.* Form, round; glands, globose; flowers, small; flesh, almost white; color, whitish, dots of red, clear red to sun; size, 2 to 1; quality, 2; season, August; freestone.

REMARKS.—New York Rareripe of Coxe.

NIVETTE. Form, roundish, slightly oval; leaves, globose; flowers, small; flesh, greenish white; color, yellowish green, red cheek; size, 1; quality, 2; season, August; freestone.

REMARKS.—A good fruit, both for the North and South. A regular bearer; a rich, delicious flavor.

NOBLESSE, or *Vanguard.* Form, round obovate; leaves,

without glands; flowers, large; flesh, greenish white; color, pale greenish white; two shades of red to sun; size, 1 to 2; quality, 1; season, August; freestone.

REMARKS.—English; of the highest character. Wherever it has been tried it has given great satisfaction. Most delicious and valuable. Tree, hardy and productive. No cultivator should be without it.—In the London Horticultural Society's Garden, at Chiswick, even as early as 1826, we find that they have at least fifty varieties of the native peaches of America—the selection from the extensive native orchards of this fruit, raised in the Middle and Western States, for distillation and other purposes. All these, many of which are so fine in our climate, and which are so grateful to travelers, as well as ourselves, with the exception of *only two*, are rejected as worthless, not being adapted to their latitude; and, owing to the want of sun and length of season, even on walls and the warmest aspects and situations to which they are obliged to confine them. It is about the same with our other native fruits, so superior in our own climate—on trial they are also obliged to reject them—the splendid and delicious apples of America—the selections of two centuries. On the other hand, there are not, comparatively, a great number of foreign fruits, particularly those of Northern latitudes, which, when brought down to our own, do not lose a great portion of that high reputation which they may have there been entitled to, compared with our native seedlings. There are, however, some foreign fruits which enjoy a latitude more like our own, as the Mela Carle, from Italy, one of the greatest apples in the world—at least it is so considered—which do very well with us, although this apple, in the climate of England, is a very ordinary fruit, and although the temperature of our climate is very considerably different from those parts of Europe, of Asia, and of Africa, in latitudes which are very similar to them.

OLD MIXON; cling. Form, round ovate; glands, globose; flowers, small; flesh, pale white, with red dots; color, yellowish white; size, 1; quality, 1; season, August.

REMARKS. — American. Deserving, with Hyslop, a place in every collection. This is one of the finest of clingstone peaches; ripens in the first part of Autumn.

OLD MIXON; free. Form, roundish oval; glands, globose; flowers, small; flesh, white, with red at stone; color, yellowish white, and pale green with dull red mingled; size, 1; quality, 1; season, September and October.

REMARKS.—Merits a place in every collection. Stands late frosts, in the Spring, better than any other, and ripens for market when most other varieties are out. All this makes it uncommonly profitable. A rather ugly, straggling grower. In selecting trees from a nursery, choose them more because of their strength and vigor of body, than for the beauty, or symmetry of their form, or the greatness of their hight. If a tree has good roots, chiefly, it will take good care of itself, and will grow according to its character. It will be sure to assume its true, natural shape. Mr. Sayers, a good nurseryman, informs us, that he is often quite glad to retain the trees that most persons reject. Though trees may have lost some of their leading branches, or are otherwise apparently deformed, when there is a strong constitution they will soon outgrow such, in reality, slight defects.

ORANGE CLINGSTONE. Form, round; leaves, no glands; flowers, small; flesh, yellow; color, deep, or dark red, red cheek at times; size, 2; quality, 2; season, August.

REMARKS.—American. Firm, juicy, vinous.

PRESIDENT. Form, roundish ovate; glands, globose;

flowers, small; flesh, white, red at stone; color, downy pale yellowish green, red cheek; size, 1 to 2; quality, 1; season, August; freestone.

REMARKS.—Ripens a little later than Morris White, or middle of August. Similar to the Rareripes, a great acquisition as a market fruit. Rich, sweet, juicy, high flavor.

PRINCE'S RED RARERIPE. See Late Red Rareripe.

RED CHEEK MALACATUNE. See Malacatune.

RODMAN'S CLING, or *Red.* Form, oblong; glands, reniform; flowers, small; flesh, white, firm, juicy; color, mostly red in sun; size, 1; quality, 1; season, August and September; clingstone.

REMARKS.—Recommended by R. Buchanan, A. H. Ernst, and J. A. Warder, to the American Pomological Society.

ROSEBANK. Form, round, compressed at apex; flesh, thick, whitish yellow, very little red about stone; color, whitish yellow, red cheek; size, 2; quality, 1; season, August.

REMARKS.—American. Trees, healthy, moderate, but regular bearers. Juicy, rich, excellent flavor.

ROYAL KENSINGTON. See Grosse Mignonne, Stetson's Seedling. Form, roundish, full at the base, and tapering a little to a very prominent point at the apex; suture, very indistinct, with a moderately deep and narrow cavity at the stem; glands, globose; flowers, small; color, slightly downy, greenish white, delicately and beautifully marbled and abruptly shaded with deep crimson on the sunny side; size, 1; quality, 1; season, August; freestone.

REMARKS.—This fine peach ranks with the Noblesse, in

size, surpassing it in beauty, and equaling it in its delicious flavor. It was an accidental seedling. It is among the very choicest in cultivation. Although, like this, many fine peaches are raised from seed, yet to rely upon this mode for the general purposes of cultivation, as Mr. Hovey remarks, is neither economy of time nor money. Budding is the most certain method, and will be the means of saving both time and money.

SWEET WATER. See Early Sweet Water.

SMOCK FREE. Form, ovate; flesh, bright yellow, red at stone; color, light orange yellow, red mottled; size, 1; quality, 1; season, October.

REMARKS.—Very late. Much esteemed for orchard culture. From New Jersey.

SNOW. Form, oval; leaves, reniform; flowers, small; flesh, white, juicy; color, thin, clear white; size, 2; quality, 2; season, August; freestone.

REMARKS.—American. Requires sunny exposures, and rich, deep soils. The blossoms of this variety are white, and the wood is a light green. Prune this tree rather low. Low-headed trees are, on many accounts, to be preferred, in our climate, even for the apple, the great orchard fruit. These should have their training commenced in the nursery; but it is seldom there attempted, on account of the desire of most purchasers to see tall trees. Often, mere whip-stalks, trimmed up clean and straight, will sell more readily than stout, stocky young trees, containing every element of future beauty and usefulness. Always select such when it is possible; remembering that the ground planted in fruit should not be appropriated to pasturage, and hence the tall stems are not needed to keep the fruit and foliage up out of the reach of cattle.

A frequent examination of their condition should be made during the growing season, and with good judgment, and small sacrifice of wood, great good may be effected. This should consist in stopping rambling, or rampant shoots, either by pinching their buds with the thumb and finger, or cutting them back with the knife: here, however, is the point to exercise great judgment. In branching the tree it should be an object from the first, to divide the head among more than two main limbs, since the division into only two is more apt to be followed by injury from splitting in after years, from the weight of the fruit and foliage, than when the strain is more divided.

St. Louis.

Remarks.—So called by the Chairman. A large yellow peach, native of this country. Chiefly valuable for its large size, and marketable qualities.—*Fruits of Missouri*, by Thomas Allen, of St. Louis.

TETON DE VENUS. Form, round, divided by furrows; flesh, red near stone; color, fine yellow, red down next sun; size, 1; quality, 1; season, August; freestone.

Remarks.—See Late Admirable; identical.

Tillotson. Form, round; leaves, glandless; flowers, small; flesh, white, red at stone; color, nearly white, red dots, dark red next sun; size, 2; quality, 1; season, July.

Remarks.—A few days before the Early York. American. Requires a good rich soil. It is nearly the earliest fruit known. Trees hardy, but not good bearers while young. Very good for early marketing. Juicy, rich, high flavor.

TIPPECANOE. Form, roundish; glands, reniform;

flowers, small; flesh, yellow; color, yellow, red in sun; size, 1; quality, 2; season, August to September.

REMARKS.—American. Juicy, sprightly. Very good.

TROTH'S EARLY.

REMARKS.—Very fine. "Excellently adapted for the locality of Cincinnati."—*F. G. Cary.*

VAN ZANDT'S SUPERB. Form, round, one side enlarged; glands, obscure; flowers, small; flesh, white, red at stone; color, white, marbled with clear red; size, 2; quality, 2; season, August.

REMARKS. — American. Not fit for market, but for small gardens for dessert. In this place it will be highly prized by amateurs.

VANGUARD. See *Noblesse.*

WALTERS' EARLY. Form, globular, flattened; glands, globose; flowers, small; flesh, white, red at stone; color, white, rich red cheek; size, 1 to 2; quality, 1; season, August; freestone.

REMARKS.—American. A variety very popular among orchardists. Adapted to light soils. Trees healthy, hardy and productive.

WASHINGTON. Form, broad, depressed, with a broad deep suture extending nearly around it; leaves, globose; flowers, small; flesh, yellowish white; color, yellowish white; size, 1; quality, 2; season, September; freestone.

REMARKS. — American. Tender, juicy, sweet. Fruit ripens late.

WHITE IMPERIAL. Form, roundish; leaves, globose;

flowers, small; flesh, mealy, white; color, yellowish white; size, 1; quality, 3; season, August; freestone.

REMARKS.—Valuable for the North. Melting, juicy; sweet, delicate, delicious flavor.

WARD'S FREESTONE. Form, roundish; flowers, small; flesh, mealy, white; color, yellowish white, red cheek when exposed to sun; size, 1 to 2; quality, 1; season, September; freestone.

REMARKS.—American. One of the very best late fruits. Trees vigorous, healthy; not too rapid growth. Juicy, vinous, and, for a late peach, sweet and delicious flavor. Invaluable for late preserving.

YELLOW RED RARERIPE. Form, round; glands, globose; flowers, small; flesh, deep yellow, red at stone; color, deep orange yellow, red dots, rich red cheek, shaded off in streaks; size, 1; quality, 2; season, July to August.

REMARKS.—American. The greatest fruit, either for market or garden. Very desirable to get it correct, as there are some spurious sorts. Juicy, melting; rich, vinous, nearly first-rate flavor.

I will close this description of peaches by remarking that in planting all kinds of trees, one important thing should be strictly observed, which is, not to plant too deep. The roots should have, as much as possible, all the beneficial influences of air and light. This is according to nature. We may observe it in those trees which spring up naturally. So much the more the crown of the plant, which may not improperly be called the seat of life, is below the surface, so much the worse. To plant this deep is to go contrary to nature. There are two different kinds of roots. One set holds the tree in the ground, and the other, the spongioles, take up or absorb the liquid nourishment. The main tap roots are in connection with the heart of the plant. In one word, it is very desirable to know what is the proper depth to plant, whether it be the seeds or the roots. In either case, we believe, no deeper than what is necessary for its protection from too much sun, and what is requisite for it to partake of proper nutriment from the soil, with moisture, and a certain degree of heat and air.

Seedling Peaches exhibited before the Horticultural Society on different occasions.

Bledsoe's Seedling, at Frankfort, Ky. Very good.

Carter's Seedling, back of Newport, Ky. Very good; freestone.

Gaddis's Seedling, back of Newport, Ky. Freestone. Good.

COOK'S SEEDLING; Nursery, Walnut Hills. Very good, and immensely productive.

Ive's Seedling. Cincinnati. Valuable.

Mottier's Seedling. Good.

BUSH'S SEEDLING. Covington. Good.

Hooper's Seedling, back of Newport. Good.

KITTREDGE'S SEEDLING. Good.

Hoffner's Seedling. Cumminsville. Very Good.

FRUITS OF OHIO.

Statement of R. Buchanan, A. H. Ernst, *and* J. A. Warder, *of Cincinnati, Hamilton county, Ohio, as reported to the American Pomological Society, at their annual meeting, held at the City of Boston, in September,* 1854.

PEACHES.

"Average bearing every other year, or one out of two or three, in favorable positions. Nearly every variety succeeds here, and our warm suns and soils have provided some splendid specimens in favorable seasons.

"The worm is kept from destroying the trees by the usual methods—picking out, and placing ashes, lime, or warm manures around the stem of the tree at the root. The latter is preferred, as the peach-tree is a great feeder, and requires manure and good culture. With these requisites, no yellows need be feared in this region. It is only necessary to give the following as a few of the varieties in general culture:

"Baltimore Rose, Coolidge's Favorite, Crawford's Early, Crawford's Late, Early York, George IV., Grosse Mignonne, Jaques' Rareripe, Late Admirable, Late Heath Cling; Morris' Red, Morris' White, New York Rareripe, Old Mixon, President, Rodman's Cling."

List of one hundred Peaches for the vicinity of Cincinnati.

Red Magdalen, 10; Early York, 5; New York Rareripe, 10; Red Cheek Malacatune, 5; Morris' Red, 5; Late Apricot Peach, 10; Baltimore Rose Cling, 5; Columbia, 5; Morris' White, 10; President, 5; Mammoth Cling, 5; Yellow Admirable, 10; Newington Cling, 5.

For Preserves.—Heath Cling, October, 5.

List of Peaches recommended by the Ohio State Pomological Society for Ohio.

Large Early York (not serrate); Crawford's Early Malacatune; George the IV.; New York Rareripe, or Morris' Red; Old Mixon Free; Old Mixon Cling; Crawford's Late; Smock Free; Heath Cling.

List of Peaches for Southern and Central Ohio, Indiana, Illinois, Southern Iowa, Northern Kentucky, Maryland, Tennessee, and Northern Missouri.

THE TWELVE BEST FOR GARDENS.

BERGEN'S YELLOW. Season, early July to early Sept.

COOLEDGE'S FAVORITE. Season, early July to last Aug.

EARLY YORK. Season, middle June to middle Aug.

GROSSE MIGNONNE. Season, early July to middle Aug.

HEATH'S CLINGSTONE (for the South). Season, middle September.

HYSLOP CLINGSTONE. Season, first Sept. to first Oct.

LARGE EARLY YORK. Season, middle July to last Aug.

MORRIS' WHITE, for preserving. Season, middle July to middle Sept.

OLD MIXON CLINGSTONE. Season, last July to early September.

OLD MIXON FREESTONE. Season, middle July to early September.

WARD'S FREESTONE. Season, middle Aug. to last Sept.

YELLOW RARERIPE. Season, early July to last Aug.

VAN ZANDT'S SUPERB. Season, early July to early Sept.

THE TWELVE BEST FOR MARKET.

ALBERGE. Season, early July to middle Aug.

CRAWFORD'S EARLY. Season, middle July to early Sept.

CRAWFORD'S LATE. Season, early Aug. to last Sept.

COLUMBIA. Season, early Aug. to early Sept.

HEATH'S CLINGSTONE (for South). Season, Sept.
LEMON CLINGSTONE. Season, early Aug. to Sept.
LARGE EARLY YORK. Season, middle July to last Aug.
PRESIDENT. Season, middle Aug. to middle Sept.
WARD'S FREESTONE. Season, early Sept. to early Oct.
YELLOW RARERIPE. Season, early July to last Aug.
WALTER'S EARLY. Season, early July to last Aug.
TROTH'S EARLY RED. Season, middle June to last July.
HYSLOP'S CLINGSTONE. Season, last Aug. to early Oct.

The following List of Peaches is presented as a guide for planting, by the Kentucky Horticultural Society.

The best varieties of each, are in Roman letters; inferior ones, ripening with them, in *italics*. The time of ripening, as given, is of much value to us here.

July 13—1st week—White Nutmeg.

July 20—2d week—Yellow Nutmeg. As yet no good fruits.

July 27—3d week—Early York, Early Anne.

August 3—4th week—Early Crawford, Early Tillotson, *President*, *Walters' Early*, *Grosse Mignonne*.

August 10—5th week—Pope's Cling, Van Zandt's Superb, Yellow Alberge, *Cole's Red*, *Haines' Early Red*, *Washington Red*, *Davis' Early*.

August 17—6th week—Hill's, Rodman's Cling, Malta, Hill's Jersey, *Royal George* (*or Teton de Venus, of some*), *Davis' Cling*, *Jacques' Rareripe*, *Late Newington*, *Spanish Cling*, Williamson's Cling.

August 24—7th week—Leopold, Catherine, Crawford's Late Orange, Free Red, Pine Apple (or Grosse Mignonne, of some), Lemon Cling, *Grosse Admirable Jaune*, *Large Malacatune*, *Green Catherine*, *Lemon Free*, *Old Mixon Cling*, *Breckenridge Cling*, and many others—(glut of peach season).

August 31—8th week—Red Rareripe (not large, but one of the best fruits for cream, or the dessert), Red Cheek

Malacatune, Pavie Pompone, *Young's Seedling*, *Hike's Seedling*.

September 3—9th week—At this period there is a defect in the present list of succession, and peaches are scarce for some two weeks.

September 10—10th week—Grand Admirable, Columbia, Whitehead's Red Heath, Smock's Late Free.

September 17—11th week—The weather is now cooler, and the same varieties last several weeks, ripening slowly.

September 24—12th week—Columbia, Lagrange, Large Heath, Freestone Heath.

October 1—13th week—Ford's Late Yellow, White's Favorite. These are the latest well-flavored peaches of the season.

N. B.—The author of this list is aware that Troth's Early, Druid Hill, "Stump of the World," and a few other varieties, rather new, have not been fruited here, and are not, therefore, classified; with this exception, the list is believed to be reliable.

PEACHES, WHICH SUCCEED AT FRANKFORT, KY.

CLINGS.—Late Heath, Old Newington, Late Newington, Breckenridge Cling (same as Old Mixon Cling), Sweet Malacatune of Spain, Baltimore Rose (the finest cling of all; it follows the Breckenridge, a local name at Louisville, and in size and beauty is every thing that can be desired; it is succeeded by the Sweet Malacatune of Spain, which is also one of the finest of its *season*), very handsome size, beautiful appearance, and very productive; the Grand Admirable.

FREESTONES.—Royal Kensington (probably Grosse Mignonne), Belle de Vitry, Walters' Early, Carpenter's Red Rareripe.

NECTARINES.

BOSTON. Leaves, globose; flowers, small; flesh, yellow to stone; color, light yellow, deep red cheek; size, 1; quality, 1; season, August; freestone.

REMARKS.—One of the best; adapted to all locations.

DOWNTON. Leaves, reniform; flowers, small; flesh, pale green; color, pale green, violet, red cheek; size, 1; quality, 1; season, July; freestone.

REMARKS.—Melting, rich, high flavor.

EARLY VIOLET. Leaves, reniform; flowers, small; flesh, white, red at stone; color, yellow, green in shade, dark purple in sun; size, 1; season, July and August; freestone.

REMARKS.—Foreign. Tree, very hardy. Not the Elruge. Melting, juicy, rich, and high-flavored.

ELRUGE. Leaves, reniform; flowers, small; flesh, green; color, pale green; size, 2; season, August; freestone.

REMARKS.—Foreign. Rich, high flavor.

We give merely the names of the following, as they are seldom cultivated here, being oftener destroyed by the curculio, even, than the plum:—Barker, Early Newington, French Yellow, Hunt's Tawny, Hardwicke, Large Early Violet, Neate's White Orange, Red Roman, Stanwicke.

From Report of R. BUCHANAN, A. H. ERNST, *and* J. A. WARDER, *to the American Pomological Society.*

NECTARINES.

"This fruit, with us, is less hardy than the peach, and is liable to be destroyed by the curculio. The varieties most in esteem, are the Early Violet, Elruge, Golden, Lewis."

APRICOTS.

Alberge. Size, 3; color, yellow; form, roundish; quality, 2; freestone; season, July and August.

Remarks.—The seed of the Alberge apricot will produce the same fruit, or with very little alteration. The Apricot tree produces its blossom buds, not only on the last year's wood, but also on the curzons, or spurs from the two-year-old wood. Great care should be used, in pruning, not to injure them; and it is advisable to remove all foreright shoots in the growing time. The Brussels and the Breda apricots make the best standards. They can all be propagated by grafting them on plum stocks.

Breda. Size, 2 to 3; color, orange brown, orange in sun; form, roundish, suture well marked; flesh, deep orange, juicy, rich, high-flavored; freestone; season, June and July.

Remarks.—This old variety is very hardy, a good grower, productive, although small, and hangs well, even after ripe. Bears late Spring frosts well.

Early Golden. Size, 2 to 3; color, pale orange; form, round ovate, narrow suture; flesh, orange, juicy, sweet; freestone; season, June and July.

Remarks.—Tree, thrifty, yet close-wooded, hardy, productive, and bears late frosts well. Valuable for market.

Large Early. Size, 2 to 1; color, pale orange in shade, brown orange in sun; form, oblate, compressed; flesh, pale orange, rich, juicy; freestone; season, June and July.

Remarks.—The best large early .apricot known, and an abundant bearer. Foliage large, leaves smaller toward the foot-stalks. Grosse Precoce of the French.

MOOR-PARK, or *Park*. Size, 1; color, orange in shade, deep orange in sun; form, roundish; flesh, firm, brown orange, juicy, rich, luscious; freestone; season, July.

REMARKS.—An old, well known, fine English variety.

We merely give the names of the following, as they are so little cultivated here:—Burlington, Brown's Early, Kaisha, Large Red, Musch, Roman, Royal, Shipley, Shaker Para, Turkey.

From Report of R. BUCHANAN, A. H. ERNST, *and* J. A. WARDER, *to the American Pomological Society.*

APRICOTS.

"Apricots bear, in sheltered situations, one out of three years. The tree flowers too early for this climate; but on walls, and in protected positions it succeeds pretty well. This fruit, like the nectarine, is only cultivated in amateur gardens.

"The favorite varieties are, the Breda, Large Early, Moorpark."

These varieties are recommended as best suited to the locality of Cincinnati.

The north sides of hills and buildings are best for apricots, to retard their blossoming, and thus tend to save the fruit from Spring frosts. In 1855, our great fruit year, apricots were abundant in our stalls, in Cincinnati. They were, chiefly, the Early Golden, and a few of one or two other kinds.

The apricot is almost always budded in the plum stock. Seedling apricots are usually more hardy and productive here than the finest grafted sorts.

PLUMS.

BLEECKER'S GAGE. Wood, downy; color, rich yellow; form, rounded oval, very regular; size, 2; stone, separating; quality, 1; use, table; flavor, sweet and luscious.

REMARKS.—From Albany, N. Y. Tree of a healthy, hardy habit, and a regular, sure bearer. Recommended by Ex-presidents Buchanan and Ernst, and the present incumbent (1857), Dr. Warder, of the Cincinnati Horticultural Society, for the locality of Cincinnati.

BLUE GAGE. Wood, downy; color, blue; form, round; size, 3; stone, free or separating; quality, 3; use, table; flavor, juicy, a little acid and somewhat rich; season, July.

REMARKS.—Foreign. Of sweet and pleasant flavor, very hardy, but the poorest of all the gages. It bears most abundant crops every season, and the seedlings make good stocks. It is the Azure Hative of the French. Recommended by Buchanan, Ernst, and Warder, for the locality of Cincinnati.

CHERRY PLUM, or EARLY SCARLET, etc. Wood, round; color, lively red, with very little bloom; form, round; size, 3; stone, adheres; quality, 2; use, table; flavor, melting, soft, very juicy, with a pleasant, lively, sub-acid; not rich nor high-tasted; season, July.

REMARKS.—On the trees they resemble cherries. Its blossoms are very thick in the Spring, which, from their earliness, renders them liable to be cut off by frost. It appears to be a native of this country. There is the common cherry plum, or *Myrobolan*, of Europe, which is rather larger, and shaped like a heart. In all else the same. There is also the *Golden.*

Cherry Plum. A seedling from the Cherry Plum, which is worthy as a market plum in Philadelphia.

Chickasaw Plum (Prunus Chicasa, Michaux).

Remarks.—Fruit about three-quarters of an inch in diameter, round, and red or yellowish red, of a pleasant sub-acid flavor; ripens pretty early; skin thin. The branches are thorny, the head rather bushy, with narrow lanceolate, serrulate leaves, looking a little distance off somewhat like those of a peach tree. It usually grows about twelve or fourteen feet high, but on the prairies of Arkansas it is only three or four feet high, and in this form it is also common in Texas. The dwarf Texas plum described by Kenrick is only this species. It is quite ornamental.

COE'S GOLDEN DROP. Wood, smooth; color, light yellow, dotted next sun; form, oval; size, 1; stone, adheres; quality, 1; use, table; flavor, rich and sweet; rather coarse-grained; season, August.

Remarks.—An English variety. Tree only moderately productive. Sometimes confounded with Yellow Egg, which is a little larger. It is one of the most delicious of all plums. It is nearly as large as the Washington. It can compare with the Green Gage (the richest-flavored of all plums, as the Seckel is of pears, and the American Golden Russet of apples) in point of lusciousness, and as it comes in after both these are gone, it may be ranked as one of the best kinds yet produced, and ought to have a place in the very smallest collection. Hovey considers it an abundant bearer. It will keep a long time after it is gathered, if placed in fine paper and kept in a dry room. Recommended by Buchanan, Ernst, and Warder, for the locality of Cincinnati.

DAMSON. Color, purple, blue bloom; form, oval; size, 3; stone, separating partially; quality, 1; use, kitchen; flavor, very juicy, melting, acid until quite mature; season, August and September.

REMARKS.—Most excellent, and the best for preserves, pies, and puddings, being rich and strong-flavored. Productive from seed. Melting and juicy; rather tart until very ripe. Good for the market and locality of Cincinnati.

DAMSON WINTER. Color, purple; form, almost round; size, 3; stone, separating partially; quality, 1; use, kitchen; flavor, acid, juicy, and rich; season, latter end of September or beginning of October.

REMARKS.—Valuable from its extreme lateness. Flesh, greenish, acid, with a slight astringency. Recommended by Buchanan, Ernst, and Warder, for the vicinity of Cincinnati.

Denniston's Superb. Wood, downy; color, pale yellowish; form, round and a little flat; size, 2 to 1; stone, separating; quality, 2; use, table; flavor, rich, vinous, and juicy; season, July.

REMARKS.—From Albany, N. Y. Of great productiveness. Not of the best quality.

DUANE'S PURPLE. Wood, downy; color, reddish purple in sun, pale red in shade, with yellow specks; form, oblong, oval, one side enlarged; size, 1; stone, adhesive; quality, 1; use, table; flavor, juicy, sprightly, moderately sweet; season, July.

REMARKS.—Foreign. Flesh, amber color, juicy, slightly acid. Recommended by Buchanan, Ernst, and Warder, for the locality of Cincinnati. In wet seasons, which are the great trouble of plums, especially, of all fruits, this plum suffers from rotting and mildew; but in warm seasons it is first rate. It is apt sometimes to be confounded with the Purple Magnum Bonum. It ripens rather early, and is a highly tempting dessert fruit.

EARLY ORLEANS. Wood, downy; color, dark reddish purple; form, round, oval; size, 2; stone, separating; quality, 1; use, table and kitchen; flavor, brisk, rich; season, July.

REMARKS.—An English variety of early maturity, and of good quality. Recommended by Buchanan, Ernst, and Warder for the locality of Cincinnati.

EMPEROR, or *Goliath*. Color, purplish red; form, round oblate; size, 1 (very large); stone, adhesive; quality, 1; use, table; season, August.

REMARKS.—Successfully raised in poultry-yards, where the fowls roost on the trees, and perambulate much all under them, making the ground hard, and frightening the Curculio away. Exhibited at the Horticultural Society Rooms by Mr. Bush, August 23, 1856. Eight bushels of plums saved in a very scarce year, raised in a yard where poultry is kept—all the fruit very fine. Mr. Bush has, also, in this way saved several other kinds.

FLUSHING GAGE. See Imperial Gage, Prince's.

GERMAN PRUNE, or *Dutch Prune*. Wood, smooth; color, very dark purple, nearly black, dusted with some blue bloom; form, long, oval; size, 2; stone, adhesive; quality, 1; use, table; flavor, quite juicy at first, but, if allowed to hang on the tree, becomes dry, rich, and sweet; season, August.

REMARKS.—A variety with many sub-varieties, the best of which is Manning's Prune; flesh, greenish. Recommended by Buchanan, Ernst, and Warder for the locality of Cincinnati.

GREEN GAGE. Wood, smooth; color, yellowish green; size, 2; stone, separating; quality, 1; use, dessert; flavor, melting, juicy, high, luscious, sprightly; season, July.

REMARKS.—Of short-jointed, slow growth, spreading and dwarfish habit. Requires a rich, warm soil, North of 42 degrees. Recommended by Buchanan, Ernst, and Warder for the locality of Cincinnati. "Excellently well adapted to the locality of Cincinnati."—*F. G. Cary.*

IMPERIAL GAGE, PRINCE'S, or *Flushing Gage.* Wood, a little downy; color, pale green, and yellowish green; form, oval; size, 1 to 2; stone, rather adhesive; quality, 1; use, table; flavor, juicy, rich, sprightly; season, August.

REMARKS.—American variety, from Prince, Long Island. A seedling from the Green Gage. Nearly as good as the parent; rather more sprightly. Soil best, light loam, dry, or even poorish. Some seasons it rots a good deal, like other plums, but it is about as sound as any in general.

It is a fine market variety; very rich stewed, in pies, or preserved. Recommended by Buchanan, Ernst, and Warder for the locality of Cincinnati. "Excellently well adapted to the locality of Cincinnati."—*F. G. Cary.*

HORSE PLUM. Wood and branches downy; color, purple in the sun, reddish on the shaded side, with blue bloom; form, oval, with a deep suture on one side; size, 2; use, table; stone, free or separates; quality, 2; flavor, rather dry and acid; season, July.

REMARKS.—American. Seedlings make good stocks for budding. It reproduces itself from seed, like some peaches, and is almost naturalized with us in our gardens; flesh, greenish yellow. Recommended by Buchanan, Ernst, and Warder for the neighborhood of Cincinnati.

HULING'S SUPERB (not Duane's Purple). Wood, downy; color, dull greenish yellow, with a pale bloom; form, round ovate; size, 1; stone, adheres partly; quality, 2; use, table; flavor, rich, brisk, excellent; season, July and August.

REMARKS.—A noble plum. It is as large as Washington. It is hardly inferior to the Green Gage. There is a little more acidity in its sweetness than in the Green Gage. It is very productive, and in all respects a very valuable fruit. Recommended by Buchanan, Ernst and Warder for the region about Cincinnati. Flesh is slightly coarse-grained, but not enough to injure it much.

JEFFERSON. Wood, nearly smooth; color, greenish yellow and golden yellow, red to sun; form, ovate; size, 1; stone, separates; quality, 2; use, table; flavor, juicy, rich; season, July.

REMARKS.—An American variety, from Judge Buel, Albany. Very productive; free from decay in a great measure. Hangs long on the tree. It takes its position very high on the list of plums. It is nearly equal, when fully ripe, to the Green Gage. As large as the Washington; it is about ten days later. It does not appear so liable as some sorts to the attacks of wasps; perhaps on account of its dark color in comparison with other plums. Recommended by Buchanan, Ernst and Warder for the neighborhood of Cincinnati.

LAWRENCE'S FAVORITE. Wood, downy; color, a dull

yellowish green; form, roundish; size, 1; stone, separating; quality, 1; use, table; flavor, rich, excellent; season, July and August.

REMARKS.—American. Tree upright, thrifty. Bears young and abundantly fruit like the Green Gage, only larger.

LOMBARD, *Bleecker's Scarlet, and Beekman's Scarlet.* Color, delicate violet red, paler in the shade, dotted with red, and dusted thinly with bloom; form, roundish oval, slightly flattened at either end; suture obscure; size, 2; stone, adheres; quality, 2; use, table; flavor, juicy, not rich, pleasant; season, July and August.

REMARKS.—American. Thrifty, healthy, hardy and very productive, and has the power of holding its fruit more uninjured than most other sorts from the attacks of that pest, the Curculio. Leaves are much crumpled. Much cultivated and highly esteemed by Mr. P. S. Bush, of Covington, Kentucky, who has raised great crops of this kind with others. Mr. Bush plants his plums in a poultry-yard, in which the fowls are continually perambulating about and roosting on the trees. By this method he succeeds in raising fine crops. Mr. Consedine has succeeded one or two years in obtaining good crops by syringing his trees with a solution of lime and sulphur, mixed together in a barrel, and applied directly after rain.

MADISON. Wood, a little downy; color, light orange, with greenish yellow; form, roundish oval; size, 2; stone, separating; quality, 1; use, table; flavor, firm, juicy, rich and sweet; season, September.

REMARKS.—From Albany, New York. Hybrid from Bleecker's Gage, and Blue Gage. Matures late, and very hardy. Has been found, by several cultivators, to suit the West very well.

McLaughlin. Wood, smooth; color, russet yellow, with a red tinge; form, roundish, flattened; size, 2 to 1; quality, 2; use, table; flavor, firm and excellent; season, July and August.

Remarks.—American. Tree thrifty, making stout, vigorous shoots of four to six feet in a season. A round, regular head. Best adapted for the North; and, therefore, not to be recommended for this locality.

Magnum Bonum, see *Yellow Egg*, *Nectarine*. Wood, smooth; color, purple; form, roundish; size, 1; stone, partly adhesive; quality, 2; use, table; flavor, a little coarse-grained, with a rich, brisk quality; season, July and August.

Remarks.—Foreign. Flesh, greenish yellow. A noble-looking fruit. Not first-rate; inferior to the Columbia. A good and regular bearer. The Peach Plum is quite different from this; it is the Prune Peche of Brittany, and superior to, and quite distinct from, the Nectarine. Many seedlings have sprung from this variety like the parent. Recommended, however, by Messrs. Buchanan, Ernst and Warder, for the locality of this region.

Old Orleans. Wood, downy; color, reddish purple; form, round; size, 2; stone, separating; quality, 1; use, table and kitchen, chiefly; flavor, sweet, mixed with acid; quality, 2; season, July and August.

Remarks.—We have in America better sorts than this old favorite in England. It is only esteemed for the kitchen here; flesh, yellowish. Recommended, however, as a cooking fruit mostly by Messrs. Buchanan, Ernst, and Warder for our locality.

OTTOMAN. Wood, slightly downy; color, greenish yellow, dark spots below; form, roundish obovate; size,

2 to 1; stone, separating; quality, 1; use, table; flavor, juicy, sweet, excellent; season, July.

REMARKS.—Foreign. Very early. Trees hardy and abundant bearers.

PEACH PLUM, or *Duane's Purple.* Color, red; form, roundish; size, 2; stone, separating; quality, 1; use, table; flavor, a little acid, pleasant.

REMARKS.—Imported from France, in April, 1820, by James Duane, of Schenectady. It was called the Apricot Plum, because the tree's name was lost. It is the Prune Peche, of Noisette. It is a little tender. Color, red; flesh, firm, green, slightly sub-acid—beautiful appearance. It parts freely from the stone, and is distinct from the plum in the nurseries of Long Island, under the name of Duane's Purple French. Recommended by Buchanan, Ernst, and Warder for this vicinity.

PURPLE EGG. Wood, smooth; color, deep red, with gray dots; form, oval; size, 1; stone, separating; quality, 2; use, table; flavor, pleasant, not high. Long and extensively grown. It is surpassed by several ripening at the same time. Flesh, greenish, coarse. Recommended by Messrs. Buchanan, Ernst, and Warder for the vicinity of the Queen City.

PURPLE FAVORITE. Wood, smooth; color, bright purple, with golden spots; form, roundish obovate; size, 2 to 1; stone, free; quality, 2; use, table; flavor, juicy, sweet, excellent; season, July to August.

REMARKS.—American. Trees dwarfish and slender, hardy, and bear when young. Good for private gardens only. Unfit for market or orcharding.

PURPLE GAGE. Wood, smooth; color, violet dots, pale

yellow; form, roundish, a little flat; size, 2; stone, separating; quality, 1; use, table; flavor, firm, sugary, high; season, August to September.

REMARKS.—Foreign. Moderate bearer, but high flavor.

PRINCE'S YELLOW GAGE. Wood, smooth; color, golden yellow, a little clouded; form, oval; size, 2 to 1; stone, separating; quality, 2; use, kitchen; flavor, sugary, rich, rather dry; season, July.

REMARKS.—American. Abundant bearer. Carries well to market. Too dry for table. Best for market.

QUETSCHE, or *German Prune*, with a great number of synonymes. Color, purple, with a thick, blue bloom; form, long oval, near two inches long, peculiarly swollen on one side, and drawn out toward the stalk. Suture distinctly marked. Flesh, firm, green, sweet, and pleasant; separates from the stone, which is flat, very long, and a little curved; season, September.

REMARKS.—Many plums are cultivated under this name. It is, therefore, rather an uncertain title. This is partly on account of its frequently coming the same from the seed. Its quality is tolerably fair for the table, but its chief use is for drying and preserving. Great bearer, and hanging long on the tree. It is very valuable, and universal in Central Europe.

RED GAGE, or *Lombard* (which see). Wood, smooth; color, bright red; form, oval, rounded; size, 3; stone, separating; quality, 1; use, table; flavor, juicy, sweet, delicious; season, July to August.

REMARKS.—American. Hardy, vigorous; small, but of the best flavor. Very productive, and free from curculio.

SAINT MARTIN'S. Wood, smooth; color, bright yellow,

bright red in sun; size, 2; stone, adheres; quality, 1; use, kitchen; flavor, juicy, rich, excellent; season, September.

REMARKS.—German. A good bearer. Profitable for market or preserving. Will hang two weeks. A kind of prune.

Sand Plum, or Beach Plum (*Prunus Maritima, Wang*).

REMARKS.—A low shrub, with stout, straggling branches, found mostly on the sandy sea coast, from Massachusetts to Virginia, and seldom ripening well elsewhere. Fruit, roundish, scarcely an inch in diameter, red or purple, covered with a bloom; pleasant, but somewhat astringent. Leaves oval, finely serrate.

SCHENECTADY. Wood, smooth; color, greenish yellow; form, round, oval, broad at stem end; size, 1 to 2; stone, separating; quality, 1; use, table; flavor, juicy, rich, sweet, delicious, melting; season. July and August.

REMARKS.—American. Tree thrifty, hardy, and prolific.

SMITH'S ORLEANS. Wood, nearly smooth; color, deep purplish red, with small golden specks, and deep blue bloom; form, ovate oblate, largest at base; size, 1; stone, adhering; quality, 2; use, kitchen; flavor, tender, juicy; season, August.

REMARKS.—American, from Long Island. Fine for market only. Flesh, yellow. Recommended by Buchanan, Ernst, and Warder for the region round Cincinnati.

THOMAS PLUM. Wood, nearly smooth; color, deep amber colored, beautifully mottled and shaded with bright red on the sunny side near the point, profusely sprinkled with white dots, and covered with a thin whitish or pale lilac bloom; size, 1; stone, freely separating; quality, 1;

use, table; juicy, sweet and pleasantly flavored; season, August.

Remarks.—Though not equal to the Washington, Imperial Gage, Jefferson, and some American plums, it possesses a combination of qualities which render it a very desirable variety. It is nearly as large as the Washington, and quite as beautiful, and hangs longer on the tree than most large plums. The trees are vigorous, early in bearing, and produce abundantly. It is very little subject to rot on the tree. For a large collection, it is a desirable plum.

WASHINGTON. Wood, downy; color, dull yellow, with little spots of red green; form, roundish oval; size, 1; stone, separating; quality, 1; use, table and kitchen; flavor, firm, sweet; season, July or August.

Remarks.—American. Uniformly hardy and productive. Very Large. Often free from curculio. Good for market at the North. Recommended by Buchanan, Ernst and Warder for the vicinity of Cincinnati. "Excellently well adapted for the locality of Cincinnati."—*F. G. Cary.*

Wild Red or Yellow Plum. (P. Americana, Marshall.)

Remarks.—Fruit, roundish oval, skin thick, reddish orange, with a juicy, yellow, sub-acid pulp. The leaves are ovate, coarsely serrate, and the old branches rough and somewhat thorny. Grows in hedges and by the banks of streams, from Canada to the Gulf of Mexico. Tree from ten to fifteen feet high. Fruit ripens in July or August.

Yellow Egg. See Egg Yellow. Wood, smooth; color, yellow; form, ovate; size, 1; stone, adheres; quality, 2; use, kitchen; flavor, not high, rather poor; season, August.

REMARKS.—Foreign. Only esteemed or cooking or market purposes. Recommended by Buchanan, Ernst, and Warder for the region around us. Its large size gives it rather an imposing appearance. It grows to a larger size in England, from its probably requiring a cooler climate than ours.

From Report of R. BUCHANAN, A. H. ERNST, *and* J. A. WARDER, *to the American Pomological Society.*

PLUMS.

"Most varieties of this fruit bear well here, when protected from the curculio; and in some seasons, when all fruits are abundant, even without protection. Average bearing, three out of four years. The curculio is destroyed by shaking it off in the morning and evening on sheets, or by syringing the tree several times with sulphur and lime water (five pounds of flower of sulphur and a half barrel of lime to a barrel of water), or by planting the trees in pavements, or in a well protected chicken-yard, apart from other fruits. The varieties most generally cultivated are as follows:

"Bleecker's Gage, Blue Gage, Coe's Golden Drop, Duane's Purple, Early Orleans, Flushing Gage, German Prune, Green Gage, House Plum, Huling's Superb, Jefferson, Nectarine, Old Orleans, Peach, Prince's Imperial, Purple Damson, Purple Egg, Smith's Orleans, Washington, Yellow Egg."

THE FIFTEEN BEST PLUMS.

The following range of seasons is from the maturity of the fruit from the northern to the southern parts of the country.

BLEECKER'S GAGE. Season, first July to last Aug.

COE'S GOLDEN DROP. Season, early Aug. to last Sept.

Early Purple, for South. Season, early June.

Early Orleans. Season, last June to first Aug.

Green Gage. Season, early July to middle Aug.

Imperial Gage. Season, first July to first Sept.

Jefferson. Season, last July to last Aug.

McLaughlin. Season, last July to last Aug.

Imperial Ottoman. Season, middle June to last July.

Purple Favorite. Season, middle July to last Aug.

Purple Gage. Season, middle Aug. to middle Sept.

Prince's Yellow Gage. Season, middle June to early August.

Red Gage. Season, July to middle Aug.

Saint Martin. Season, last Aug. to first Oct.

Washington. Season, July to last Aug.

Smith's Orleans. Season, July to last Aug.

A list of Plums recommended by Henry Ward Beecher *for Indiana.*

Green Gage, Jefferson, Huling's Superb, Coe's Golden Drop, Purple Gage, Cruger's Scarlet, Washington, Red Gage or *Lombard*, Smith's Orleans, Royal de Tours.

The following are suitable for light, sandy soils, on which the curculio makes the greatest ravages, and on which plums usually drop their fruit: Cruger's Scarlet, Imperial Gage, Red Gage or Lombard, Coe's Golden Drop, Bleecker's Gage, Blue Gage.

CHERRIES.

Adam's Crown. Color, pale red; form, round, heart-shaped; size, 2; use, table; quality, 2; flesh, tender, juicy, agreeable; season, June.

Remarks.—Of English origin. Tree vigorous. Class of sweet cherry.—The cherry is rather short-lived. It does not generally continue more than thirty years in perfection.—It is observed of stone fruit in general, that if sown immediately after they are excarnated, they will appear the following Spring, but being kept too long, they will not germinate under two years.—The cherry-tree produces its fruit, generally, at the extremity of the branches; therefore, in pruning, they should never be shortened.

American Heart. Color, pale yellow and red; form, heart-shaped, compressed; size, 2; use, table; quality, 2; flesh, yellowish, juicy, sweet; season, May and June.

Remarks.—Tree vigorous. Spreading.

Anne. Size, 2; use, table; quality, 1; flesh, excellent flavor.

Remarks.—Sweet-cherry class.

Archduke. Form, heart-shaped, compressed; size, 1; use, table; quality, 2, flesh, light red, tender, sub-acid.

Remarks.—Not May Duke.

BELLE DE CHOISY. Color, red; form, round; size, 1; use, table; quality, 1; flesh, tender, sweet; season, June.

Remarks.—Very handsome and good. Recommended for this reason by Buchanan, Ernst, and Warder, successively Presidents of the Cincinnati Horticultural Society, and eminent as practical pomologists.

Belle of Orleans. Color, yellowish white; form, roundish; size, 2; use, table; quality, 1; flesh, tender, juicy, and delicious; season, May and June.

Remarks.—Promises to be valuable.

Belle Magnifique. Color, clear rich red on pale yellow; form, ovate rounded, heart-shaped; size, 2 to 1; use, table; quality, 2; flesh, tender, sub-acid; season, June to August.

Remarks.—Tree of Duke habit. Hardy and vigorous. Moderate bearer. Fruit hangs well.

BIGARREAU MOTTLED. Color, amber with red; form, rounded, heart-shaped; size, 1; use, table; flesh, yellowish white; season, June.

Remarks.—Of American origin. Tree vigorous, very productive. Half tender, juicy, sweet. Not high flavor. Recommended by Buchanan, Ernst, and Warder for the locality of Cincinnati.

BIGARREAU NAPOLEON. See Napoleon Bigarreau.

BLACK EAGLE. Form, obtuse, heart-shaped; size, 1 to 2; use, table; quality, 2; flesh, reddish purple, half tender, rich, sweet flavor; season, June.

Remarks.—From England; by the great botanist and pomologist, Charles Knight. Not very productive. Fruit borne in threes. One of the finest cherries in cultivation. No collection can be considered complete without it. "Excellently well adapted for Cincinnati."—*F. G. Cary*, late

President of the Cincinnati Horticultural Society, and a good practical fruit cultivator, at Farmers' College, near Cincinnati.

BLACK TARTARIAN. Color, purplish black; form, heart-shaped; size, 1; use, table; quality, 2; flesh, half tender, juicy, sweet, mild, pleasant, not high flavor; season, June.

REMARKS.—Of peculiar, upright growth. Requires more pruning than others, or it will become too dense. Recommended by Buchanan, Ernst and Warder for the vicinity of Cincinnati. "Excellently well adapted for the locality of Cincinnati."—*F. G. Cary.*

BLACK HAWK. Color, dark purplish black; form, heart-shaped, often obtuse; size, 1; use, table; quality, 1; flesh, almost firm, rich, sweet, fine flavor; season, June and July.

REMARKS.—Raised by Professor Kirtland, Cleveland, Ohio, the judicious originator of many fine and valuable cherries and other good fruits, and possessing and cultivating much science in horticulture, etc. This cherry is also among the fruits recommended by Buchanan, Ernst and Warder for the region of Cincinnati.

BRANT. Color, reddish black; form, heart-shaped, rounded angular; size, 1; use, table; quality, 2; flesh, tender, juicy, sweet and rich.

REMARKS.—Raised by Professor Kirtland. Tree vigorous. Flowers irregularly.

BURR'S SEEDLING. Color, clear bright red; form, heart-shaped; size, 2 to 1; use, table; quality, 2; flesh, half tender, juicy, sweet; season, May and June.

REMARKS.—New York origin. Productive, vigorous.

Buttner's Yellow. Color, pale yellow; form, regular, heart-shaped; size, 2; use, table; quality, 3; flesh, whitish yellow, rather tough; season, July.

Remarks.—From Germany.

Carnation. Color, yellowish white; form, roundish; size, 1 to 2; use, table; quality, 2; flesh, tender, juicy, and not acid when fully ripe; season, June and July.

Remarks.—A variety of Morello. Hardy.

China Bigarreau. Color, bright amber yellow; form, roundish, heart-shaped; size, 2; use, table; quality, 2; flesh, half tender, juicy and rich; a little bitter till ripe; season, June.

Remarks.—Very productive.

Cleveland. Color, bright clear red, on amber yellow ground; form, round, heart-shaped, flattened at apex; size, 1; use, table; quality, 2; flesh, juicy, rich, fine flavor, pale yellowish white, almost firm; season, June.

Remarks.—One of Professor Kirtland's raising in 1842. Tree thrifty and very productive.

Coe's Transparent. Color, pale light amber yellow, with bright clear red; form, regular round, slightly angular at junction of stem; size, 2; use, table; quality, 2; flesh, tender, juicy, rich; season, June and July.

Remarks.—Tree vigorous; healthy habit.

Delicate. Color, rich amber yellow, mottled; form, regular, roundish flattened, slight suture on side; size, 2 to 1; quality, 1; flesh, rich, juicy, sweet, high flavor, delicate, translucent (a pretty quality); season, June.

Remarks.—Raised by Professor Kirtland. Tree thrifty, productive and healthy.

DOCTOR. Color, light yellow and red; form, roundish, heart-shaped, with a suture extending all around; size, 2; use, table; quality, 1; flesh, juicy, tender, sweet, with a delicious flavor; season, June.

REMARKS.—One of Professor Kirtland's. Not very vigorous, but all his seedling cherries are of a healthy habit. Bearing too great crops, although a fault on the right side.

DOWNER'S LATE. Color, bright lively red, mottled amber in shade; form, round, heart-shaped, slightly compressed on one side, size, 1; use, table; quality, 2; flesh, tender, delicious, and sweet when fully ripe; season, June and July.

REMARKS.—Tree vigorous, hardy, prolific and healthy. Recommended by Buchanan, Ernst and Warder for the vicinity of Cincinnati.

DOWNING'S RED CHEEK. Color, yellowish white; form, obtuse, heart-shaped; size, 2 to 1; use, table; quality, 2; flesh, half tender, and delicately sweet; season, July and August.

EARLY PURPLE GUIGNE. Form, roundish heart-shaped, indenture at point; use, table; quality, 1; flesh, juicy, rich, sweet, and excellent; season, May and June.

REMARKS.—By some called German May Duke. Only moderately productive when young. It is very early, and delicious. It is highly deserving of cultivation, being the earliest yet known; more so, even, than the May Duke, and Early May, with the same advantages of situation and soil. The May Duke is quite green, and the Early May is hardly ripe, when the Early Guigne is in full perfection. It is nearly two weeks earlier than the May Duke, and fully equal to it in quality. The Early May hangs so long, that it will come up to the Guigne at last. This cherry is even earlier than the May Bigarreau.

EARLY PROLIFIC. Color, bright carmine red, mottled on light amber yellow; form, round obtuse, heart-shaped; size, 1; use, table; quality, 1; flesh, half tender, almost firm, juicy, rich, sweet, and delicate flavor; season, May and June.

REMARKS.—Professor Kirtland. Tree, healthy, vigorous, and upright. Recommended by Buchanan, Ernst, and Warder, for the locality of Cincinnati.

EARLY RICHMOND, *Pie Cherry, Kentish, Early May.* Color, bright red, becoming darker as it ripens; form, round; size, 2; use, chiefly kitchen, very ripe for table, though rather too acid; quality, 1; flesh, juicy, very tender, sprightly, rich, acid flavor; season, May and June.

REMARKS.—Very early, valuable, and hardy. Of Morello family. Excellent for early market, for stewing, pies, etc. It also has the fine quality of hanging long. The Early May is so very like the above, that the distinction is hardly worth making. Both very productive, and "excellent for Cincinnati."—*F. G. Cary.* Recommended, also, as a matter of course, by those good judges, Buchanan, Ernst, and Warder, for the locality of Cincinnati.

ELTON. Color, shining pale yellow on the shaded side, but with a cheek next the sun delicately mottled and streaked with bright red; form, long, heart-shaped, pointed; size, 1; use, table; quality, 1; flesh, nearly tender, juicy, sweet, with an exceedingly rich, high flavor; season, June.

REMARKS.—English. Of superior quality. Trees, grow vigorously, with a rather drooping habit. Recommended by Messrs. Buchanan, Ernst, and Warder, for this vicinity.

GOVERNOR WOOD. Color, clear rich red; form, roundish, heart-shaped; size, 1; use, table; quality, 1; flesh, light pale yellow, half juicy, tender, sweet, and fine.

REMARKS.—Raised by Professor Kirtland. Tree, a vigorous, healthy grower (like most of his seedlings, which, constitutes much of their value, beside their qualities), productive while young, very remarkably good. Recommended by Messrs. Buchanan, Ernst, and Warder, for the locality of Cincinnati.

GRAFFION. See Yellow Spanish.

HILDESHEIM. Color, yellow, mottled with red; form, heart-shaped; size, 2; use, table; quality, 1; flesh, pale yellow, firm, sweet; season, July and August.

REMARKS.—From Germany. Tree, upright, strong grower. Unproductive while young. Bears late.

HORTENSE. Color, bright lively red, mottled on amber; form, round, elongated side, compressed; size, 1; use, table and kitchen; quality, 1; flesh, rich, sprightly, subacid; season, July.

REMARKS.—From France. Of Duke habit. Vigorous and healthy. Suited to rich, moist soils. Moderately prolific.

JOC-O-SOT. Color, dark liver; firm, very regular, uniform, heart-shaped, slightly obtuse, deep indenture at apex; size, 1; use, table; quality, 2; flesh, dark liver, tender, juicy, rich, sweet flavor; season, June.

REMARKS.—Kirtland. Tree thrifty, round-headed, and productive.

KIRTLAND'S MARY. Color, light and dark rich red; form, roundish, heart-shaped; size, 1; use, table;

quality, 1; flesh, light yellow, firm, rich, juicy, sweet, and very highly flavored; season, early in June.

REMARKS.—Raised by the Professor. Very beautiful, very desirable, and early, for both family and market. Recommended by Buchanan, Ernst and Warder for the vicinity of Cincinnati. Took premium in Spring of 1855, when exhibited by the author at the Cincinnati Horticultural Society, as the best earliest cherry. Tree handsome, rather upright.

KIRTLAND'S MAMMOTH. Color, light clear yellow; form, obtuse, heart-shaped; size, 1; use, table; quality, 2; flesh, almost tender, juicy, sweet, high flavor; season, June.

REMARKS.—Another seedling of the Professor. Tree, large and vigorous. Only moderately productive.

LATE BIGARREAU. Color, rich yellow ground, red cheek; form, obtuse, heart-shaped, broad indenture at the apex; size, 1; use, table; quality, 2; flesh, firm, juicy, sweet, of agreeable flavor; season, June and July.

REMARKS.—Dr. Kirtland. Tree, vigorous. and very productive.

LARGE HEART-SHAPED. Color, dark shining red; form, roundish, heart-shaped; size, 1; use, table and kitchen; quality, 2 to 3; flesh, coarse tissue, not juicy, nor high-flavored; season, last of June.

REMARKS.—From France. Tree, strong and vigorous; very productive, and desirable for market.

LATE DUKE. Color, rich deep shining red, when mature; form, roundish, heart-shaped, with a slight suture on one side; size, 1; use, table; quality, 1; season, beginning of July to the end.

Remarks.—From France, although of English origin. Every late cherry must prove a valuable acquisition to this fine fruit. Of this character is the Late Duke. It is of very large size, of a beautiful color, and an abundant bearer. It is equal to the May Duke. The fruit of the Late Duke is usually borne in pairs, or threes, on a short stem, about a quarter of an inch in length.

Logan. Color, liver; form, obtuse, sometimes regular, heart-shaped; size, 2 to 1; use, table; quality, 2; flesh, nearly firm, juicy, rich, sweet flavor; season, June.

Remarks.—Kirtland. Tree, hardy and healthy, moderately productive. Little liable to injury by frost.

MARY. See Kirtland's.

May Bigarreau (*Bigarreau de Mai*, *Beauman's May*, *or Allen's Favorite*). Color, rich deep red, when fully mature becoming of a shining, dark, purplish color; form, oval, heart-shaped; size, 2 to 3 (rather small); flesh, purplish red, soft, and tender, juice abundant, with a sweet, rich flavor; quality, 1; season, last of May, sometimes the beginning of May, or early part of June.

Remarks.—With the exception of the Early Purple Guigne, which is not yet much known, the May Bigarreau may be safely set down as at least two weeks earlier than any variety; unless perhaps the Early May, or Early Richmond, may equal it in this respect.

MAY DUKE. Color, deep red; form, roundish obtuse, heart-shaped; size, 1; use, table; quality, 2 to 1; flesh, reddish, tender, sub-acid; season, May to June.

Remarks.—From France. Of a hardy, upright growth; produces freely, but ripens very irregularly. Good for the Cincinnati market. "Excellently well adapted to the locality of Cincinnati."—*F. G. Cary.*

MOTTLED BIGARREAU. See Bigarreau Mottled.

NAPOLEON BIGARREAU. Color, pale yellow, becoming amber in the shade, richly dotted, and spotted with very deep red, and with a fine marbled, dark crimson cheek; form, roundish, obtuse, heart-shaped, with a suture line frequently raised, instead of being depressed; size, 1; use, table and kitchen; quality, 1; flesh, very firm, moderately juicy; season, June.

REMARKS.—Tree, vigorous, productive. Showy for market. Good for cooking. "Excellently well adapted for the vicinity of Cincinnati."—*F. G. Cary.*

OSCEOLA. Color, dark purplish red; form, regular, heart-shaped; size, 2 to 1; use, table; quality, 1; flesh, juicy, rich, and sweet; season, June.

REMARKS.—One of those many good seedlings which Dr. Kirtland had the singular judgment and good fortune to raise by selection, etc. Tree round, hardy, and healthy. A good bearer.

OX HEART, or WHITE BIGARREAU. Yellowish red in sun; form, heart-shaped; size, 1; use, table and kitchen; quality, 2; flesh, almost firm; season, June.

REMARKS.—Sweet, delicious. Not a very good bearer.

PONTIAC. Color, dark purplish red; form, obtuse, heart-shaped; size, 1; use, table; quality, 2; flesh, juicy, sweet, and agreeable.

REMARKS.—From Professor Kirtland. Tree vigorous and healthy. Recommended by Messrs. Buchanan, Ernst, and Warder, for the locality of Cincinnati.

POWHATTAN. Color, rich purplish red; form, roundish flattened; size, 2; use, table; quality, 2; flesh, half tender, juicy, sweet, pleasant; not high flavor; season, late June to July.

REMARKS.—Kirtland's. Tree vigorous and productive. Profitable for a market fruit. Late, and regular in size.

RED JACKET. Form, regular, obtuse, lengthened; size, 1; use, table; quality, 2; flesh, half tender, juicy, of good, not high flavor; season, June to July.

REMARKS.—Another of Professor Kirtland's. Fruit ripens late. Excellent for market. Recommended by Buchanan, Ernst, and Warder, for the region of Ohio.

ROCKPORT. Color, brilliant, deep red; form, obtuse, heart-shaped; size, 1; use, table; flesh, firm, juicy, sweet, rich, and delicious; season, June, early.

REMARKS.—Professor Kirtland's raising. One of the hardiest of his fine seedling cherries, as was fully proven in the memorable severe cold of last Winter (1856). A few of these cherries were found rather too tender in that great and unusual trial. But it is not very likely we shall soon have another such a winter. The wood was also not matured enough to meet it. Tree strong, vigorous, upright habit. Valuable for private market gardens. Recommended by Buchanan, Ernst, and Warder, for the region of Ohio.

SHANNON. Globular, flat at junction with stem; size, 1 to 2; use, kitchen; quality, 2; flesh, juicy, acid; season, June.

REMARKS.—Still another of the Professor's. A Morello. Tree very hardy.

SWEET MONTMORENCY. Color, pale amber in the

shade, of a deep orange red in the sun, becoming darker when fully ripe, and mottled with yellow; form, nearly round, little flattened at both ends, with a shallow suture on one side, and an indented point at the apex; size, 2 to 3; use, table; quality, 1; season, July.

REMARKS.—The number of American varieties of fruits, and cherries in particular, is yearly increasing; and we may soon expect to find the principal kinds, in general cultivation, our native varieties. Mr. Knight produced several new sorts of cherries by cross-fertilization, which have stood high; and Dr. Kirtland's success speaks well for itself. The Sweet Montmorency is the production of Mr. Manning, in this country. It is an accidental seedling in 1831 or 1832.• It was produced in Mr. J. F. Allen's garden in Salem. It hardly ever fails to ripen a crop of fruit every season. It is scarcely ever injured by weather (particularly if wet), which usually cracks and injures most varieties. It is one of the latest sweet cherries, ripening at the same time as the May Duke, and hangs long after it is mature, and also keeping some time after it is gathered. It is very productive, and bears sometimes 100 cherries on a small forked branch a foot long.

TECUMSEH. Color, reddish purple; form, obtuse, heart-shaped; size, 2 to 1; use, table and kitchen; quality, 2; flesh, sweet, juicy, but not high-flavored; season, July.

REMARKS.—Kirtland's again. Tree moderately vigorous. Hardy, late, for market.

YELLOW SPANISH. See Graffion. Color, whitish yellow, with mottled red in the sun; form, regular obtuse, heart-shaped; size, 1; use, table; flesh, yellowish; form, juicy, rich, sweet, delicious; season, June.

REMARKS.—One of the richest and best cherries. But

has rather a tendency to decay. Tree strong, spreading, healthy, and productive. Recommended by Messrs. Buchanan, Ernst, and Warder, for the locality of Cincinnati. "Excellently well adapted for the locality of Cincinnati." —*F. G. Cary.*

From Report of R. Buchanan, A. H. Ernst, *and* J. A. Warder, *to the American Pomological Society.*

CHERRIES.

"Cherries bear, on an average, one out of three years. The climate of Southern Ohio is too warm for this fruit, and but few varieties succeed well here. The best cherry region in our State is the southern shore of Lake Erie, where fine crops are produced almost every year. The rose-bug and the slug, there complained of, do not annoy us here; but the trees of the finer varieties often crack open in winter, after warm wet Autumns, and are either destroyed or greatly disfigured.

"The western country is largely indebted to Dr. J. P. Kirtland, of Cleveland, for the production of some very fine seedling cherries, better adapted to the climate than those of foreign origin. The hardiest varieties with us are of the 'Morello' family; next are the 'Dukes,' and least of all the 'Bigarreaus.'

"The following are mostly cultivated:

"Belle de Choisy, Black Hawk, Black Tartarian, Downer's Late Red, Early May, Early Prolific, Elton, Governor Wood, Kirtland's Mammoth, Kirtland's Mary, Reine Hortense, Carnation, May Duke, Mottled Bigarreau, Napoleon, Pontiac, Red Jacket, Rockport (hardiest, very early), White Bigarreau, Yellow Spanish.

Dwarf cherry trees are produced by propagating the Sweet or Duke varieties on the Mahaleb or Morello roots. They should be worked just at the crown of the root.

Merits of Cherries, decided upon at a Cherry Festival held at Cleveland, Ohio; the best situation (on the Lake) and soil for Cherries.

KIRTLAND'S MARY. Large, pale red, firm, first quality; a very pretty cherry.

GOVERNOR WOOD. New. A large, round, pale red, sweet, first quality; a delicious cherry; one of the very best.

VIRGINIA MAY DUKE. Small, heart-shaped, bright red, second-rate.

ROCKPORT BIGARREAU. Large, handsome, bright red, first-rate.

DAVENPORT'S EARLY BLACK. Large, soft, black, sweet; very good.

CLEVELAND BIGARREAU. Large, bright red, first quality; a great bearer.

LOUIS PHILLIPE. New; Morello; large, dark red, tart, rich.

GRIDLEY. Small, dark, second-rate.

DOWNTON. Pale red, solid, rich; good.

BELLE DE CHOISY. Round, red, soft, rich; first rate.

MADISON BIGARREAU. Medium size, bright red, soft, sweet; second-rate.

MANNING'S MOTTLED BIGARREAU. Pale red; second-rate.

CHINA BIGARREAU. Small, pale red; poor.

EARLY WHITE HEART. Medium, pale red, sweet, firm; good.

BOYER'S EARLY. Medium, etc., as above.

ROBERT'S RED. Medium. These three are very similar, and suspected to be the same.

DOCTOR. Above medium, red, firm, rich; first quality.

JUNE DUKE. A tart variety of May Duke.

KNIGHT'S EARLY BLACK. Dark, rich, soft; first quality. Said to be a poor grower and poor bearer.

BLACK HEART. This cherry is smaller than it should be, and appears to be only a good mazzard, but it is said to improve as it ripens, and to bear well.

ENGLISH AMBER. Not nearly equal to its American namesake.

SWEDISH. This is the Rockport Bigarreau.

DELICATE. A new cherry, of a pale but bright and delicate color. Size, moderate; quality, good.

ELIZABETH. Above medium, bright color, flavor brisk; good.

BLACK OX-HEART. Dark, and not large.

OHIO BEAUTY. Very handsome, good, and a great bearer.

BLACK HAWK. Medium size, dark red; a superior fruit, becoming a liver-colored black when ripe; excellent for market.

MAMMOTH. Large, light red, tender, very fine; one of the best, but not so good a bearer.

OSCEOLA. Good size, black, very pleasant; handsome on the tree, and a good bearer.

RED JACKET. Medium size, light red, good, and a great bearer.

ELLIOTT'S FAVORITE. Very handsome on the tree; a great bearer, and very hardy.

JOCKOTOS. Large, black, tender, pleasant, and a great bearer.

THE TWELVE BEST SWEET CHERRIES.

All cherries ripen at the South (that is, with us) about one month earlier than the date here fixed. These are all good market varieties, with the exception of "Delicate," and "Early Purple Guigne."

BELLE OF ORLEANS. Use, table; season, early June.

BRANT. Use, table; season, middle June.

BLACK TARTARIAN. Use, table; season, last June.

Black Hawk. Use, table; season, last June.

Coe's Transparent. Use, table; season, last June.

Delicate. Use, table; season, first July.

Downer's Late. Use, table; season, middle July.

Early Purple Guigne. Use, table; season, first June.

Elton. Use, table; season, last June.

Governor Wood. Use, table; season, middle June.

Kirtland's Mary. Use, table; season, last June.

Rockport. Use, table; season, middle June.

THE SIX BEST DUKE CHERRIES.

Archduke. Use, table and cooking; season, early July.

Belle de Choisy. Use, table; season, last June.

Belle Magnifique. Use, cooking; season, July and August.

May Duke. Use, table; season, June.

Reine Hortense. Use, table; season, middle July.

Vail's August Duke. Use, table and cooking; season, August.

THE SIX BEST MORELLO CHERRIES.

Carnation (for South and West). Use, table and cooking; season, July.

Early Richmond. Use, cooking; season, June.

Shannon. Use, table and cooking; season, middle July.

Donna Maria. Use, cooking; season, middle July.

Imperial. Use, Cooking; season, August.

Louis Phillipe. Use, cooking; season, middle July.

The "Large Morello," originated by Professor Kirtland, will probably supersede Carnation, but it is not yet sufficiently tested.

QUINCES.

Common Quince.

Remarks.—This and the two following are often confounded with each other. It is probable that from seeds of either sort, varieties have been, and still may be obtained, some of which would produce apple-shaped, and some pear-shaped fruit.

Apple-Shaped Quince.

Pear-Shaped Quince.

Remarks.—Medium size; roundish oblong, or pyriform, tapering to the stalk; skin, dull yellow; flesh, firm, tough, dry, but of high flavor. When stewed, or cooked, it is less tender, and the flesh less lively in color, than the Orange Quince. Leaves, oblong, ovate; season, September.

Portugal Quince.

Remarks.—Very good, and distinct from the preceding sorts. It does not, however, become, except in very favorable seasons, of so deep an orange; its leaves are broader, and its growth less contracted; consequently it is used for grafting pears on. Shy bearer. Fruit, medium to large, regular oblate, pyriform, smooth; flesh, mild; cooks tender. Ripens ten days earlier than the Orange Quince.

Orange Quince, *Angers Quince.*

Remarks.—These two last enumerated varieties, possess characters differing so little from what may be found among sub-varieties of the others, that they are scarcely worth distinguishing. Excellent for dwarf pears. Leaf, round, and downy underneath.

LARGE-FRUITED QUINCE.

REMARKS.—This variety is most esteemed. Should not be gathered early. Leaf, ovate, pointed; fruit, large, ovate, oblate pyriform; skin, smooth, of a rich golden yellow.

NEW UPRIGHT QUINCE.

REMARKS.—From Ellwanger & Barry, of Rochester, New York. It grows upright, strikes readily from cuttings; but after the first year's growth seems to lose vigor, and afterward grows very tardy. Not fit to graft pears on.

Angers, Paris, or Orleans quinces, are the best stocks for dwarf pears. The Middle and Western States seem to be the peculiar home for the quince. There are four kinds, only, used for cooking. In most cases, quinces will produce the same from seed, but they will sport a little — hence the variety, although not many, of forms. They are very easily propagated from cuttings. The bush form is the most natural.

GRAPES.

ADA. Bunch, compact; color, very dark; skin, thin; flavor, sweet and vinous, very juicy; quality, 2; situation, south.

REMARKS.—Originator, Dr. Valk, of Flushing, Long Island. Strong and vigorous growth; shoots, partially brown; joints, six inches from eye to eye; leaf, large, and handsome. Fruits freely. Perfectly hardy

BLAND. Bunch, long and loose; color, pale red; skin, thin; flavor, delicate, pleasant, sweet, a little astringent; quality, 1; situation, south-east.

REMARKS.—From Virginia. It is a good table grape, where it will ripen, which is not north of Philadelphia. Late in ripening, and valuable to put away for Winter use.

CATAWBA. Bunch, medium sized, shouldered; color, pale red; form of berries, nearly round; skin, thick; flavor, slightly pulpy, sweet, juicy, rich, aromatic, musky flavor; quality, 1; situation, south-east.

REMARKS.—Highly—most highly esteemed, for dessert and wine use. In growth and foliage it resembles the Isabella, except that the wood is of darker color, shorter jointed, and more round, and at base of every leaf, there is a white downy spot. Sweet when only half ripe, but very luscious when quite ripe, and dark colored. Berries covered with a beautiful lilac bloom. Pond's Seedling, To-Kalon, Clermont, White Catawba, and the Zane, are sub-varieties of the Catawba, but not equaling it. The

"Mammoth Catawba" is also a sub-variety which, under high culture, surpasses the original, only in size.—That part of the United States between the thirty-eighth and forty-fourth parallels of latitude, so far, is entitled to the supremacy in grape culture. Already the wines of Ohio and Missouri, begin to supplant the imported Rhine and Champagne wines here, even at the same prices. Terraces rise above terraces on the hill-sides of the Ohio, and the red bluffs begin to disappear beneath masses of vine foliage, and purple clusters of fruit. We find that Indiana, Illinois, and Michigan, are improving the hint given by Ohio; in fact Indiana must be recognized as one of the pioneers, for Vevay first commenced it in the beginning of the present century. Missouri already ventures to contest the palm with Ohio.—The Catawba is twenty to one in cultivation, in Ohio, over the Isabella. Of these two grapes the best wines are made in Ohio. There is a peculiarity of these wines, that no spurious compound can be made to imitate them; and in purity and delicacy, and we may almost add richness, there is no known wine to equal them. The cuttings of these vines are always saleable, to propagate new vineyards.

Clinton. Bunch, medium or small, compact, not shouldered; color, blue bloom; form of berries, nearly round, small; skin, thin; flavor, pulpy, rather harsh; quality, 2; situation, South.

Remarks.—From Western New York. Not a strong grower, although perfectly hardy and suited to border planting, three feet apart, and stake training not exceeding four feet high, in gardens. Its greatest recommendation is that it ripens ten days or two weeks earlier than the Isabella, and is, therefore, suited to higher or more Northern latitudes.—Grapes should be extensively raised from seed, but not from the seed of foreign grapes, as has been recommended. We should extensively plant the seed of our native grapes, and great changes will be produced. Mr. Longworth had an evidence of this, he having presented at the exhibition of our Horticultural Society, a seedling from the Isabella, that passed as the Black Hamburgh. The berry was larger than the largest Black Hamburgh on the tables, though the latter was raised under glass. In raising plants from the seed of the Catawba, grapes of great value may be produced, and varieties without number. But the greater portion will go back to the origin of all, the Fox Grape. The Fox Grape is readily distinguished by the extreme white color of the leaf on the under side, and when a full blooded Fox, the stem will be covered with a heavy down. When there is a down, the plant should be thrown aside as soon as the stem shows it. When the stem is smooth, if the leaf is white, it may produce a good fruit. When the under side of the leaf has less, or not more of the Fox character than the Catawba, a plant of value may be expected. It is not the sweetest grapes that contain the most sugar. The size of the berry is not important for wine, but the vine should be of vigorous growth, and bear a good crop. Solid wood of last year's growth, two eyes to each graft, are enough; cut one inch above the upper eye, and three inches below the lower eye, it will make a cutting that will vegetate.

DIANA. Bunch, small, below medium, compact; color, pale red; form of berries, round; flavor, little pulpy, rich.

REMARKS.—Seedling from the Catawba; grown by Mrs. Diana Crehore, Boston, Massachusetts. Matures two weeks earlier than the Catawba. It has not equaled its parent at Cincinnati. It suits best a Northern latitude. Resembles the Catawba a little in flavor. It is hardy, vigorous and productive.

CONCORD.

REMARKS.—A little larger, and six days earlier than the Isabella. Very hardy, and a free grower—nearly as good as the Isabella. Suited to the North rather than for us West.

DELAWARE, *or Traminer*. Color, pale reddish; form, roundish oval; bunches, medium sized; berry, middle size, uniform, tender, juicy, sweet, without pulp, and rich and agreeable flavor.

REMARKS.—Messrs. Prince, Grant, Downing, Brinckle, Hovey, and others, consider this the most delicious native grape, except, perhaps, the Scuppernong of the South. It is very hardy. It was discovered in New Jersey, and was introduced into Ohio twenty-five or thirty years ago. This variety may be recommended as promising very well. Ripens ten days before the Isabella.

ELSINBURGH. Bunch, medium, loose shouldered; color, black; form of berries, small, round; skin, thin, blue bloom; flavor, melting, sweet; quality, 1; situation, S.E.

REMARKS.—From Salem county, N. J. A nice little grape, suited for the dessert, and for growing on the trellises in gardens. A moderate regular bearer. Ripening a little before, or with the Isabella. "Best."

HERBEMONT. Bunch, large, compact shouldered; color, purple; form of berries, small, round; skin, thin, purple bloom; flavor, sweet, excellent, juicy, vinous; quality, 1; situation, South.

REMARKS.—Of doubtful origin. Hardy. In fruit it does not differ from the Lenoir; but in wood, distinct. Only moderately vigorous.

IMITATION HAMBURGH. Bunch, large; color, dark purple; skin, thin; flavor, juicy, soft; quality, 2; situation, south.

REMARKS.—A native variety, inferior to the Black Hamburgh. Origin not known.

ISABELLA. Bunch, large, rather loose shouldered; color, dark purple; form of berries, oval, large; skin, thin; flavor, juicy, sweet, rich, a little musky aroma; quality, 1; situation, south-east.

REMARKS.—Origin somewhat disputed. Probably from South Carolina, and, therefore, a native. Its vigor and product with us, will ever render it a favorite. Berries, when fully ripe, nearly black, and then very sweet. Berries covered with a blue bloom. Hyde's Eliza, Troy Grape, Pennsylvania, Maicon, Sherman, Chillicothe Seedling, and Lee's, are all sub-varieties; not equal to the original.

KITTREDGE SEEDLING. Color, brown; form of berries, round ovate; skin, thick; flavor, rich and sweet; quality, 2; situation, south.

REMARKS.—This may become a good grape for wine.

LENOIR. Quality, 1; situation, south.

REMARKS.—More vigorous than the Herbermont, but otherwise like it. Wood, light colored, with a light blue cast.

MINOR'S SEEDLING, *or Benango.*

REMARKS.—Has value as a wine grape.

MISSOURI. Bunch, below medium, loose; color, black almost; form, small, round; flavor, tender, sweet and juicy, little pulp; quality, 2; situation, south-east.

REMARKS.—From Missouri. Of slow growth, short-jointed, and, like the Clinton, suited to border culture. A wine is made from it resembling Madeira.

MORIN.

NAUMKEAG.

NORTON'S VIRGINIA. Bunch, long, little shouldered, compact; color, deep purple; form of berries, small round; flavor, pulpy, harsh; quality, 3; situation, south-east.

REMARKS.—A native. Of but little value, but tolerably passable for the table.

OHIO SEGAR BOX. Bunch, large, loose shouldered; color, nearly black; form of berries, small round; flavor, without pulp, sweet; quality, 1; situation, south-east.

REMARKS.—True origin unknown. North of Cincinnati it fails. Only for table use.

REBECCA.

REMARKS.—No doubt a variety of the Chasselas family. The only white native grape within our knowledge, and desirable, if only for that reason. It promises well. Propably rather tender.

RULANDER.

REMARKS.—A German grape of this name, gives great promise of success in the open air. The Muscadine and Scuppernong scarcely succeed. *Fruits of Missouri*, by Thomas Allen, of St. Louis.

SCHUYLKILL, etc. Bunch, not shouldered; color, black; form of berries, large, round ovate; flavor, pulpy, juicy, firm, musky, often harsh; situation, south.

REMARKS.—From Pennsylvania. Leaves, downy.

SHAKER'S SEEDLING. See Union Village Grape.

REMARKS.—Valuable and pleasant, Southern species. There highly esteemed.

SCUPPERNONG.

UNION VILLAGE GRAPE.

REMARKS.—As large as the Black Hamburgh, and quite hardy. It is a monstrous grower, bunches quite large, the flavor sweet, and as good as the Isabella. A fine table grape. A little earlier than the Isabella and Catawba, but not so early as the Delaware. Vigorous. Probably not well adapted for making wine. Flavor very fine. Cane very stout.

FOREIGN GRAPES.

BLACK HAMBURGH. Bunch, large, shouldered both sides; color, bright purple, purple blue when ripe; form of berries, very large, roundish, oval; skin, thin; flavor, sugary, rich; quality, 1; situation, cold house, vinery.

REMARKS.—The best for the vinery. In sheltered locations, out of doors. In many cities south, as far as Cincinnati, it does well with Winter protection. A good bearer. A vine of this variety, at Hampton Court Palace, planted in 1769, produced two thousand bunches—over one ton of fruit, which the author saw.

BLACK PRINCE. Bunch, long, often shouldered; color, black, blue bloom; form of berries, large, thinly set, oval; skin, thick; flavor, sweet, excellent, very good; quality, 1; situation, cold house, vinery.

REMARKS.—Succeeds well, with Winter protection, out of doors. It hangs long in the house after fully ripe. A profuse bearer.

BLACK FRONTIGNAC. Bunch, long; color, black; form

of berries, medium size, round; skin, thin; flavor, good, musky, rich; quality, 1.

REMARKS.—Muscadine wine is made from this. A profuse bearer.

BLACK CLUSTER. Bunch, small, compact; color, black; form of berries, medium, roundish ovate; flavor, juicy, sweet; quality, 2; situation, cold house.

REMARKS.—This variety is hardy, and succeeds out of doors.

GRIZZLY FRONTIGNAN. Bunch, rather long, narrow; color, green red; form of berries, medium, round, thick bloom; flavor, juicy, rich, musky, high flavor; quality, 1; situation, cold house.

REMARKS.—Adapted only to the house. It ripens early. Best quality.

ROYAL MUSCADINE. Bunch, large, long shouldered; color, green white, and blue when ripe; form of berries, above medium, round; flavor, tender, rich, delicious; season, September; quality, 2; situation, cold house.

REMARKS.—Highly esteemed. Stronger in growth, and larger in berries than the White Sweetwater. Requires, out of doors, Winter protection, and plenty of wood ashes.

MUSCAT OF ALEXANDRIA. Bunch, very large, loose, irregular; color, pale amber; form of berries, large, oval; skin, thick; flavor, musky, rich, perfumed flavor; quality, 2; situation, hot house.

REMARKS.—Adapted only to house-culture, and benefited by artificial heat. It is the Malaga grape. Brought to this country in jars.

TRAMINER. Bunch, medium, compact; color, pale red; form of berries, round, ovate, middle size, uniform; flavor, tender, juicy, sweet, no pulp, rich, and pleasant; quality, 1.

REMARKS.—This deserves a place in every garden. Ripens ten days before the Isabella.

WHITE FRONTIGNAN. Bunch, medium size, rather long, rarely shouldered; form of berries, middle size, round, rather closely set; flavor, delicious, sugary, rich, musky flavor; quality, 1; situation, cold house.

REMARKS.—An old productive variety. Suited only to the house.

SEEDLING GRAPE.

RINTZ'S SEEDLING. Bunch, compact, medium size; color, dark purple, almost blue when fully ripe; form of berries, round, larger than Catawba, medium size.

REMARKS.—From the seed of the Catawba, by Sebastian Rintz. In wood, leaf, and habit, like the Catawba. Same vigorous growth as the Fox Grape, the parent of the Catawba. The skin is thick, pulp tough, juice sweet, but slightly astringent, and, as in the Fox grape, not abundant. Ripens one month before the Catawba. Considered by the Committee the best seedling from the Catawba raised in this vicinity; but as a wine grape, can not compare with the Catawba.

STRAWBERRIES.

ALICE MAUDE. Flowers, hermaphrodite; form, conical; size, 2; quality, 2; color, dark crimson.

REMARKS.—In some parts of Virginia extensively cultivated.

AJAX. Flowers, staminate; form, globular; size, 2; quality, 2; color, dark crimson.

REMARKS.—New.

BRITISH QUEEN. See Myatt's.

REMARKS.—Very large, and of rich flavor, but does not fruit well here.

Bicton Pine. Flowers, staminate; form, long ovate; size, 1; quality, 1; color, white; bright blush on cheek.

REMARKS.—The blossoms will not produce as much as many kinds. It should be tested by amateurs.

Bishop's Orange. Flowers, pistillate; size, 2; quality, 2; color, light orange scarlet.

REMARKS.—Moderately prolific. Fruit in clusters. Desirable in warm, deep, sandy soils.

Black Prince. Flowers, pistillate; form, round; size, 2; quality, 2; color, deep purplish red.

REMARKS.—English. Flesh, rich, red, sweet. Requires a rich, loamy soil. Of high flavor. Succeeds best North. Is there of high flavor. Is variable in quality; some seasons first-rate. It is hardy and prolific.

Boston Pine. Flowers, staminate; form, roundish, slightly conical; size, 2; quality, 2; color, deep, rich, shining red; season, early.

REMARKS.—American. Requires high cultivation, in hills; vines, vigorous; firm, juicy, sweet, with a sprightly, agreeable flavor. Productive. We will here observe that the strawberry is as easily raised from seed as any other plant, and with the certainty of producing very good varieties. Mr. Hovey states that the French cultivators raise the Alpine Strawberry in this way, as an annual, the plants bearing a fine crop the first year. To combine the greatest number of good qualities in any fruit is the great object in the growth of new kinds: the possession of a portion of them without the others must fail to give any variety a high rank for general cultivation; therefore it is a saving of both time and money to reject all of those that do not come up very near to this standard. Hovey's great seedling was a successful hit, and has at least combined a great many most precious characteristics of what a strawberry should possess. The Boston Pine has not by any means reached the elevation of its great predecessor; still, under certain circumstances of climate, soil, management, etc., it may be pronounced a respectably good fruit. In the first place, Mr. Hovey says: "It should receive good cultivation to have the fruit in fine condition. If the plants are allowed to run together, the produce will not be half a crop. The soil should be good, and there should be a space of at least a foot between the rows. Each plant throws up from six to ten stems, and if the roots do not find sufficient nourishment, many of the berries will not fill up and attain their proper size. Well grown, the plants are literally covered with fruit." We have no account of its having had any considerable success in our Western soils. It is ripe a week before Hovey's seedling, at the same time as the Old Scarlet, or Early Virginia, and continues a long time in bearing. There are good qualities in any location where it may suit. Flesh, pale scarlet, fine grained, buttery, and solid, very juicy, sweet, and rich, with a brisk, high, and delicious flavor.

British Queen. See Myatt's British Queen.

BURR'S NEW PINE. Flowers, pistillate, rather large for the sex; firm, obovate, or round; size, 2; quality, 1; color, light pale red; season, very early.

Remarks.—Origin at Columbus, Ohio, on a clayey soil, in 1846. Vines, hardy, vigorous, and productive; flesh, whitish pink. Of delicate, aromatic flavor, sweet and delicious. Too tender for a market fruit, but highly desirable in a garden. Recommended by Messrs. Buchanan, Ernst, and Warder, for the locality of Cincinnati.

Burr's Seedling. Flowers, hermaphrodite; form, roundish ovate, often conical; size, 2; quality, 2; color, light pale red.

Remarks.—American. Vines vigorous and hardy; tender, mild and pleasant. Does not bear carriage well. Valuable as a fertilizer of other kinds.

Brewer's Emperor. Flowers, staminate; form, oval; size, 2; quality, 2; color, dark red.

Remarks.—English. Hardy; said to be productive. Not much disseminated.

Brilliant. Flowers, hermaphrodite; form, conical; size, 2; quality, 2; color, deep crimson.

Remarks.—American. Flavor good; productive; plants vigorous. Described by W. R. Prince, in *Horticulturist.*

CALEB COPE. Flowers, pistillate; form, pointed; size, 2; quality, 2; color, scarlet.

REMARKS.—American. Flavor good. Prolific.

CHARLOTTE. Flowers, pistillate; form, obovate; size, 2; quality, 2; color, dark scarlet.

REMARKS.—American. Sweet, sprightly flavor. Productive. Described by W. R. Prince in *Horticulturist.*

CLEVELAND. Flowers, hermaphrodite; form, cockscomb to conical, irregular; size, 2; quality, 2; color, dark purplish red in sun, opposite clear vermilion.

REMARKS.—American. Firm; of Pineapple flavor; rich and delicious.

Crescent Seedling.

REMARKS.—From New Orleans. A perpetual bearer. Requires testing.

Climax. Flowers, pistillate; form, conical, a little necked; size, 2; quality, 3; color, light scarlet.

REMARKS.—American. Rather acid; very productive.

Cornucopia. Flowers, pistillate; form, conical; quality, 3; color, scarlet.

REMARKS.—American. Productive. Described by W. R. Prince in *Horticulturist.* Mr. Prince has flourished out in too many kinds for all to be very valuable. It is better to have *one* really great, than to have one hundred merely passably good. Without this the very best NAMES will be lavished upon them almost in vain. Better to have a wonderfully fine fruit with a plain name, than one hundred only moderately good with the most high-sounding titles.

CRIMSON CONE. Flowers, pistillate; size, 2; quality, 2.

REMARKS.—Dutch berry. Good flavor; a little acid. Vines vigorous, requiring space; productive. A great New York berry. Its defects are its second-rate size, and acid flavor.

CUSHING. Flowers hermaphrodite; form, obtuse conical; size, 2; quality, 2; color, scarlet.

REMARKS.—American; by Dr. Brinckle, Philadelphia. Flesh, fine; flavor, sprightly, agreeable; productive.

Duchesse de Trevise. Form, ovate; size, 2; quality, 3; color, deep red.

REMARKS.—Not known here. Much praised in England and France. A very different thing for us here. They describe it as having a brisk, rich flavor, and juicy, and a good bearer.

DUKE OF KENT. Flowers, staminate; form, roundish, conical; size, 3; color, bright scarlet; season, very early, ripe 1st June, or even middle of May.

REMARKS.—English. Sharp, rather acid flavor; vines, hardy.

Duncan's Seedling. Size, 2; quality, 2; color, dark rich red.

REMARKS.—English. New. Fine flavor. Productive. Lately introduced.

DUNDEE. Flowers, pistillate; form, round, ovate, very uniform; size, 2; quality, 2; color, light pale clear scarlet; season, ten days after Willey, or Hudson.

REMARKS.—Scotch. Firm, rich acid, high flavor, very productive. Great for market. Rather late. Vines very hardy.

EBERLEIN'S SEEDLING. Flowers, hermaphrodite; form, conical, compact; size, 2; quality, 2; color, dark scarlet; season, early.

REMARKS.—American. Vines vigorous. Moderately productive. Good. Slightly acid.

ELTON. Flowers, staminate; form, ovate; size, 2; quality, nearly 1; color, light red; season, very late, and valuable only on that account.

REMARKS.—English. Rather too acid. Of good size, but rather shy in bearing. Rather tender.

EXTRA RED. See McAvoy's Extra Red.

GENESSEE. Flowers, hermaphrodite; form, round; quality, 1; size, 1; color, dark crimson; season, a little late.

REMARKS.—American. Very productive. Of very fine rich, very sweet flavor. A delightful, and very desirable sort.

Green Strawberry. Form, round; size, 3.

REMARKS.—Only curious.

HAUTBOY. See Prolific Hautboy.

HOVEY'S SEEDLING. Flowers, pistillate, small; form, roundish ovate, a little conical, with a short neck, never cockscomb-shaped even in the largest berries; size, 1 (very large, commonly three to four inches in circumference); quality, 1; color, dark rich shining red, paler when grown in the shade; seeds, dark, and imbedded in a small cavity; flesh, scarlet, firm, nearly solid, abounding with a most agreeable acid, and exceedingly delicious and high-flavored juice (Burr's New Pine, and McAvoy's superior, only, surpassing it in richness); season, May; ripe, about a week after the Boston Pine, and continues in perfection during the whole strawberry season.

REMARKS.—American. Originated in 1834. Vines, very vigorous, more so than most other varieties, perfectly hardy, forming numerous runners, though seldom too many; leaves, large; leaflets, roundish, generally convex, obtusely serrated with about twenty serratures; surface, rather smooth, deep brilliant glossy green, and rarely ever spotted with brown; petioles, short; leaf-stalks, upright, medium length, moderately strong; flowers, rather small, very regular in form; petals, roundish, slightly imbricated and cupped; stamens, very short and imperfect, deficient in anthers; calyx, very small, finely divided, and quite reflexed; scapes, moderately strong, about the same length as the leaf-stalks, elevating the fruit from the ground; peduncles, rather long and slender. Every flower, when properly fertilized, is succeeded by a perfect berry. Flesh, firm, bears carrying remarkably well, of a very agreeable, sweet, lively flavor. Best in rich loam, and wood soils; impregnated with the Old Hudson male, yields immense crops. Not so good on sandy soils. In clay much better. Good for both market and amateur cultivation. Berries, very large. Sometimes, in the West, even eight inches in circumference. Suited to nearly all soils and climates. Stands drouth wonderfully well. Hardy also in Winter. This berry should be well ripened to be eaten in perfection. Recommended by Messrs. Buchanan, Ernst, and Warder, for the locality of Cincinnati. In speaking of this fine strawberry, it may be well to observe, that Mr. Keen, of Isleworth, near London, about the year 1820, made the first really great improvement in this delicious fruit, in the production of the variety so well known as Keen's Seedling. A few years ago Mr. Myatt, of Deptford, near London, succeeded in

25

raising some varieties, as the British Queen, and others, which have done wonders in England, but in our climate, our American seedlings have completely surpassed them. Mr. Hovey raised his seedlings from the seeds of Keen's Seedling, and some others, all English.

Hooper's Seedling. Flowers, staminate; form, conical; size, 2; color, dark rich red; quality, 2; season, late.

Remarks.—Of good flavor, and productive.

HUDSON, or *Hudson's Bay, Late Scarlet, American Scarlet.* Flowers, pistillate; form, ovate, often with neck; size, 2 to 1; color, rich dark, glossy red; season, May to June.

Remarks.—Most extensively cultivated, particularly round Cincinnati. Hardy, and rather late. Fine and rich, but of rather acid flavor. Excellent for preserving and for market; firm in carriage. In the new, fresh wood soils in this neighborhood it produces great crops and large fruit; not, however, quite so sure a crop as the Hovey; yet not much difference. Should hang until fully ripe. Recommended by Messrs. Buchanan, Ernst, and Warder, for the locality of Cincinnati.

IOWA MALE, or *Washington.* Flowers, staminate; form, roundish conical; size, 2 to 1; quality, 1 to 2; color, pale red; season, quite early, before the Hovey and Hudson, etc.

Remarks.—American. Delicate and good, and peculiar in flavor. A good impregnator. Immense crops of this most productive fruit are raised by the field cultivators in Kentucky, back of Newport, and are brought to Cincinnati market. Most valuable for its earliness.

Jenny's Seedling. Flowers, pistillate; form, very

regular, roundish conical; size, 2 to 1; quality, 2; color, rich glossy dark red; season, one week after Longworth's Prolific. Rather late.

REMARKS.—Firm texture; desirable for preserving. Good for general cultivation; vines very hardy. Very productive; 3,200 gathered from less than three-fourths of an acre. Rich, sub-acid, delicious; almost never failing a crop. Recommended by Messrs. Buchanan, Ernst, and Warder, for the vicinity of Cincinnati.

Jenny Lind. Flowers, pistillate; form, conical, perfect, often short neck; size, 2 to 1; quality, 2 to 1; color, bright light scarlet.

REMARKS.—Rather solid, heavy; tender and juicy; flavor, pleasant, sub-acid, and sometimes highly perfumed.

KEEN'S SEEDLING. From Indiana. Flowers, pistillate; form, round; size, 2; quality, 2; color, light crimson; season, medium.

REMARKS.—At A. H. Ernst's. Greatly productive. A great market fruit; very uniform in shape. Rather acid, but of pleasant flavor.

LARGE EARLY SCARLET. Flowers, hermaphrodite; form, roundish ovate; size, 2; quality, 2; color, bright scarlet; season, very early.

REMARKS.—Good impregnator of pistillates. Rich, slightly acid. Good market berry, on account of earliness.

LARGE WHITE BICTON PINE. See Bicton Pine.

LA GRANGE. See Prolific Hautboy.

LONGWORTH'S PROLIFIC, or *Schneicke's Seedling.* Flowers, hermaphrodite; regular roundish, or obovate;

size, 1; quality, 1; color, rich dark crimson; season, medium, with the Hudson.

REMARKS.—Cincinnati, 1848. At the Garden of Eden, by Schneicke. For market culture likely to be valuable. It is immensely productive, and its own impregnator. More firm than McAvoy's Superior, and equally large, but not so rich and good in flavor. Sub-acid. Recommended by Messrs. Buchanan, Ernst, and Warder for the locality of Cincinnati.

McAVOY'S SUPERIOR. Flowers, pistillate; form, varying, irregular, roundish, conical, sometimes a little necked; size, 1; quality, 1; color, rich, dark, glossy crimson; season, medium.

REMARKS.—Originated at Cincinnati, in 1848, on loamy soil. Received prize of $100 from Cincinnati Horticultural Society, in 1851. Tender, juicy, rich, with fine, high flavor. Too tender for long distances to market. Desirable for private gardens, and markets near town. Requires very strong and numerous fertilization—nearly plant for plant. Too tender and delicate in texture to keep and preserve its flavor so long as many kinds. It is not considered equal to Burr's New Pine in flavor, but as fine as any other.

METHVEN SCARLET. Flower, Pistillate; form, round, coxcomb; size, 2; quality, 2; color, dull scarlet; season, four or five days after general strawberry season.

REMARKS.—Scotch. Strong grower. Sometimes produces large crops.

McAVOY'S No. 1, or *Extra Red.* Flowers, pistillate; form, round, uniform; size, 1; quality, 1; color, scarlet.

REMARKS.—Originated in Cincinnati, in 1848. Flavor agreeable, rather acid. Immensely productive. Likely to become a good market fruit. Has not a very high flavor. Recommended by Messrs. Buchanan, Ernst, and Warder, for the locality of Cincinnati. Vastly hardy.

PEABODY'S NEW SEEDLING. Flowers, hermaphrodite; color, rich, deep crimson; form, irregular, and somewhat compressed; beautiful, attached to the calyx by a polished coral-like neck without seeds; size, 1 (of the largest, measuring six and seven inches in circumference); use, table; quality (not yet known here); flavor and flesh, firm, melting, and juicy; season (unknown here).

REMARKS.—It is said by the proprietor and originator that the fruit is borne on tall foot-stalks, is of the most exquisite fine flavor, and bears transportation better than any strawberry ever yet cultivated. He further states that as a proof of the keeping qualities of this new strawberry, on the morning of the 9th of May, he packed a case of the berries, took them to Columbus, six miles, in his buggy, sent them from Columbus to Savannah, 300 miles, by railroad, and from Savannah to New York, 900 miles, by steamer, to Messrs. Thorburn & Co. Mr. Thorburn stated that they came to hand on Tuesday, sound and in very good condition, retaining an unusually strong strawberry aroma. They observed that their dark color gives them a richer look, approximating to the English Hautbois, grown at New York. They added that the berries had wilted down only a very little up to that time, Friday morning, 16th May. This new seedling has been produced by crossing the Ross Phœnix with a wild strawberry of Alabama. Being hermaphrodite, it requires, of course, no impregnation, and is said to be a capital impregnator for pistillate varieties; a hardy, vigorous grower, withstanding both cold and heat without injury. In good soil, the vine is said to grow to an enormous size; single plants can not be covered by a half-bushel measure.

The berry has few seeds. It requires no sugar for the dessert, rivaling the far-famed Burr's New Pine. Prolific, opening its blossoms during the mild days of Winter, and perfecting its fruit as soon in the Spring as the weather will permit. (This description applies to Georgia.) The plant is reported very beautiful when it is in flower and leaf. If it should not succeed here as well as in Georgia, or not at all, it may be valuable to cross others with. It has been thought by some by its standing the great jolting of so long a journey, that it may be *too firm to eat well.* But this, of course, is merely conjecturing about it. So, also, although this strawberry is of the Hautbois and Pine family, and they have hitherto been of a kind not to succeed well in this country, and, therefore, very few of them are sold, in comparison with other strawberries, it does not follow that the Peabody seedling may fail also in this respect:—this also by way of suggestion.

MONROE SCARLET. Flowers, pistillate; form, roundish, short neck; size, 2; quality, 1; color, light scarlet.

REMARKS.—American. Very prolific. (Ellwanger & Barry, 1850). Surpassing most others in productiveness. It is a hybrid of Hovey's Seedling and the Duke of Kent. Fruit beautiful and good for market use, and a long bearer. Does well partially shaded.

MOYAMENSING. Flowers, pistillate; form, roundish conical; size, 2; quality, 2; color, deep crimson.

REMARKS.—American origin. Flesh, red; flavor, fine. It bore off the premium of the Pennsylvania Horticultural Society, in 1848, for the best seedling strawberry exhibited that year. Good as a market fruit.

MYATT'S BRITISH QUEEN. Flowers staminate; form, roundish; size, 1; quality, 3; color, scarlet.

REMARKS.—Raised by Mr. Myatt, Debtford, near London. Valuable in England, but in our climate infinitely inferior to our own seedlings. Flavor, rich. A poor bearer.

NECKED PINE. Flowers, pistillate; form, conical, always with neck; size, 2 to 3; quality, 1; color, light scarlet; season, medium.

REMARKS.—American. A little acid, pleasant when fully ripe. Only suited to private gardens, being tender but very productive. Recommended by Buchanan, Ernst, and Warder, for the vicinity of Cincinnati.

No. 1, or McAVOY'S EXTRA RED. See McAvoy's Extra Red, or No. 1. Stands drought and frost very well.

PENNSYLVANIA. Flowers, pistillate; form, broadly conical; size, 2; quality, 2; color, dark crimson.

REMARKS.—American. "Flavor fine."—*Trans. Pennsylvania Horticultural Society.*

PROLIFIC HAUTBOY, or LA GRANGE. Form, round conical; size, 2; quality, 2; color, dark purplish red.

REMARKS.—The only high wood class worth cultivation. Very musky in flavor.

Red Alpine. Flowers, perfect; form, conical; size, 3; quality, 3; color, bright scarlet.

REMARKS.—Of a delicate and peculiar flavor. It ripens gradually a long time—its chief value. The White Alpine varies only in color from it. Destroy early blossoms, and a full crop may be had.

RIVAL HUDSON. Flowers, pistillate; size, 2; quality, 2; color, bright scarlet.

REMARKS.—Flesh, red, firm, sub-acid. Very productive. Much like Hudson, or Hudson Bay, its parent. Originated the same time as Burr's New Pine.

ROSS PHŒNIX. Flowers, staminate; form, round, conical; size, 2; quality, 2 to 3; color, dark red.
REMARKS.—American. Generally a poor bearer. Firm, and of good flavor.

Ruby. Form, ovate; size, 2; quality, 2; color, ruby red.
REMARKS.—English. Good flavor; prolific.

Southborough. Flowers, pistillate; form, ovate, conical; size, 2; quality, 2; color, rich deep scarlet; season, early.
REMARKS.—A good mate to the Early Scarlet. Fruits at same time. Vines hardy.

Swainstone Seedling. Flowers, staminate; form, ovate; size, 2; quality, 2; color, light crimson.
REMARKS.—Ripens a long time, but an uncertain bearer.

WALKER'S SEEDLING. Flowers, staminate; form, roundish conical; size, 2; quality, 1; color, very dark crimson; season, medium.
REMARKS.—Worthy of attention. Very hardy. American. Of sprightly, rich flavor. Prolific for a staminate. A great bearer. Described by Col. Wilder in *Horticulturist.*

WESTERN QUEEN. Flowers, pistillate; form, regular, round conical; size, 2 to 1; quality, 2; color, rich, dark, glossy red; season, medium.
REMARKS.—Origin, Cleveland; by Professor Kirtland, in 1849. Firm, juicy, sub-acid, sprightly, and of agreeable flavor. Bears carriage well; considered by many better than the Hudson.

RASPBERRIES.

ALLEN.

REMARKS.—A native variety; extensively raised at Black Rock, New York; equal to most foreign kinds in use. It is a reddish black. It is indispensable with the Black Cap and Ohio Everbearing in any collection.

ANTWERP RED. Color, red, dull; quality, 1.

REMARKS.—Rarely found true, West. It is a Dutch sort. It is regularly long-conical. A rich, sweet flavor. Canes moderately strong, yellowish green, becoming pale brown early in Autumn—nearly smooth in the upper portions. Ripens from June to July. It requires protection in Winter, by forking the canes down and covering with earth or litter. Though the American kinds require less trouble in protection in Winter, etc., yet they do not much diminish the value of the foreign kinds, as they ripen at different periods. Near large cities this berry is among the best, as it always yields fair crops.

ANTWERP YELLOW. Color, yellow; quality, 1.

REMARKS.—Suited to small gardens. It sometimes throws up a succession of shoots, maturing fruit for a long time. Shoots, strong, light yellow, with greenish spines. Requires protection in Winter. The Antwerps do not suit the Southern States. The fruit is rather long in shape, very tender, rich and delicate. All the Antwerps, to produce good crops, need to be covered in Winter. It is a great trouble and expense. Probably some of the native kinds, with common Black Cap, if improved by better cultivation, would be found the best suited to our wants. They, at present at least, sell well in market. The Antwerps require to be picked every day, while the Black Cap remains on longer, and, therefore, can be more easily gathered, as it is much more firm

American Red. Color, red; quality, 3.

Remarks.—Very common. Shoots, upright, light brown. Fruit, medium, roundish, light red, sub-acid and tender. Early—in June and July. Grows wild in some places.

American White. Color, white; quality, 3.

Remarks.—Like the above. More firm than the Red. Very sweet, but of little flavor.

AMERICAN BLACK, *Thimble Berry, or Black Cap.* Color, black; quality, 2.

Remarks.—Grows wild. It increases much in size in a rich garden soil; and, ripening late, is profitable for market. It is liked for jam, puddings, etc. It is a firm berry. Shoots, long, rambling, recurved; berries, dark purple, nearly black, round, flattened. It will yield large crops, and will probably give as much satisfaction as any other kind, perhaps more. It is very hardy. The foreign kinds are rather too tender and unproductive, but richer.

Bannet. Color, red, quality, 2.

Remarks.—English. Shoots, long, yellowish green, branching; fruit, large, soft, roundish conical, purplish red; agreeable.

BLACK RASPBERRY. See *American Black.*

Brentford Cane. Color, dull red; quality, 3.

Remarks.—English. Shoots, strong, branching; fruit, medium, oval, conical. Inferior.

COPE. Color, crimson; quality, 2.

REMARKS.—A seedling of Dr. Brinckle's. Not much disseminated. Foliage, light green; fruit, large, conical, crimson; spines, red.

COLONEL WILDER. Color, cream; quality, 1.

REMARKS.—Dr. Brinckle's seedling; which is hardy, firm, light colored, and good for market. Shoots, strong, light colored, and very hardy; fruit, above medium, roundish conical, light cream color, a sprightly, fine flavor. Productive, and ripening its fruit in succession, early to late in season. Raised from seed of the Fastolf.

CUSHING. Color, cream; quality, 2.

REMARKS.—Raised by Dr. Brinckle from seed of the Double-bearing. Not much known, but deserving attention of amateurs. Shoots, strong, vigorous; prickles, brown; leaf, plaited, regular, firm. Fruit, large, roundish conical, crimson, and of fine flavor. Matures early; said to produce sometimes, when the season is wet, a second crop in Autumn.

CORNWALL'S RED. See Bannet.

Cox's Honey. Color, yellowish white; quality, 3.

REMARKS.—English. Fruit, medium, borne in clusters along the stems.

COMMON RED. See American Red.

COMMON BLACK CUP. See American Black.

Creton Red. Color, red; quality, 2.

REMARKS.—From the Mediterranean. Shoots, upright, hardy. Fruit, medium, round, deep red, acid. Late.

DOUBLE-BEARING (*Late Cane*). Color, red; quality, 2.

REMARKS.—Variety of the Antwerp. Large, dull red, hardy. Ripens late. Perpetual bearing (which means late in the Autumn, as well as in the Summer).

EMILY. Color, light yellow; quality, 2.

REMARKS.—From Col. Wilder. Little disseminated; vigorous growth, with white spines. Fruit, large, round, occasionally shouldered; light yellow.

EVER-BEARING OHIO. Color, blackish purple; quality, 2.

REMARKS.—A valuable, very hardy, early berry, for amateurs. Produces a good crop from the shoots of the same year in the Fall, if the weather is moist and favorable. Carries well to market—a rich, pleasant berry. The ends of the shoots can be easily made to take root by inserting them in the ground. Native of the northern part of Ohio. Cultivated at the Quakers' Settlement. Introduced in Cincinnati by N. Longworth. A valuable addition to a collection.

FRENCH. Color, cream; quality, 2.

REMARKS.—From Dr. Brinckle. From the Fastolf. Fruit, large, round, crimson, matures late; spines, red.

FULTON. From French seed. Fruit, large, round, crimson; productive, and vigorous grower; spines, red.

FASTOLF. Color, bright purplish red.

REMARKS.—English. Worthy of praise; though, like the Antwerp and some others, tender, and requiring protection in Winter. Less tender than the R. Antwerp. Shoots, strong, and much inclined to branch; light yellowish brown. Fruit, large, roundish, obtuse conical, soft, rich, high flavored, productive; ripening its fruit in long continued succession.

FRANCONIA. Color, dark rich red.

REMARKS.—From France. Its canes are nearly hardy, but in most Winters require protection, maturing good crops. Shoots, strong, branching, yellowish brown, with scattered, rather stout bristles. Leaves, rather narrow. Fruit, large, obtuse conical, dark rich red, rather acid flavor; more firm than Fastolf, and not so much as R. Antwerp. A few days later than the latter in ripening.

Gen. Patterson. Color, crimson; quality, 2.

REMARKS.—From Col. Wilder. Vigorous grower. Spines red. Fruit, large, round, crimson.

KNEVET'S GIANT. Color, deep red.

REMARKS.—A good English fruit, very early. Canes very strong, and nearly hardy. Fruit very large, conical, of excellent flavor.

MONTHLY (*Large Fruited Monthly, etc.*). Color, red; quality, 2.

REMARKS.—Excellent and productive; requiring a peculiar system of culture, like most others of this kind, to produce fruit in succession. Shoots, long, slender, purplish in the sun, thickly covered with dark purple spines. Fruit, hardly above medium, fine flavor. To produce an Autumn crop, prune the canes in the Spring to within a foot of the ground.

MRS. WILDER. Color, cream; quality, 2.

REMARKS.—Seedling of Col. Wilder. Nearly resembles the Col. Wilder.

Nottingham Scarlet. Color, scarlet; quality, 2.
REMARKS.—English. Fruit, medium; obtuse conical.

ORANGE. Color, brown orange; quality, 1
REMARKS.—From Dr. Brinckle, in 1844. Nearly hardy, and rather late. Good for market. Shoots, vigorous, with white spines; leaf, irregular. Fruit, large, ovate, and of excellent flavor. Productive.

OHIO. See Ever-bearing.

VICTORIA. Color, red; quality, 2.
REMARKS.—Medium, roundish conical.

WALKER. Color, deep crimson; quality, 2.
REMARKS.—From Dr. Brinckle. Promises well for market. Fruit, large, round, solid; adheres firmly to the stem, keeps long in perfection on the plant, and bears carriage well; spines, red.

WHITE THIMBLE BERRY. Color, whitish yellow; quality, 2. Very like the American Red, except in color.

Woodward's Red Globe. Color, red; quality, 2.
REMARKS.—English. Large red, roundish conical.

WILMOT'S EARLY RED. Color, red; quality, 2.
REMARKS.—English. Small, roundish, red. Early.

BLACKBERRIES.

High Bush.

Low Bush.

LAWTON, or *New Rochelle.* Color, black; quality, 1.

Remarks.—Very large, a great bearer, pulpy and delicious flavor. It loves a cool, moist, shady soil, is easily cultivated, and is everywhere becoming a universal favorite. Adopted by the Congress of Fruit Growers. The "New Rochelle Blackberry" sends up, annually, large and vigorous, upright shoots, with lateral branches, all of which, under common cultivation, will be crowded with fine fruit; a portion ripens daily, in most seasons, for six weeks, commencing about the first of July. They are perfectly hardy, always thrifty and productive, and have not been found liable to blight, or injury by insects. To produce berries of the very largest, they should have a heavy, damp soil, and shade. This will be a good starting point for seedlings. It is a most valuable improvement in this fruit.

CURRANTS.

American Black.

Remarks.—Not very productive—better shaded. We will here observe, that this valuable variety is seldom cultivated as it should be—that it bears transportation to market, without injury, that it grows well in all soils, and under almost any mode of culture; that with extra attention, and manure, the Common White and Red Dutch, yield fine crops.

BLACK NAPLES. Color, black; quality, 1.

REMARKS.—The largest fruited, and most productive of the black currants. Hardy in all sections of the United States, but in the South not productive. Requires shade, and a rich soil. Burns up in a very sunny exposure.

BLACK ENGLISH. Color, black; quality. 3.

REMARKS.—Not productive here—bunches too short.

BLACK GRAPE. Color, black; quality, 2.

REMARKS.—Better than the above. Makes excellent wine, and good for medicinal purposes.

CHAMPAGNE. Color, pale red; quality, 3. Very acid.

CHERRY. Color, red.

REMARKS.—Very high flavor. Not very productive.

GONDOUIN. Color, red; quality, 1.

REMARKS.—From France. Strong growth, large, matures late; should be further tested, but not largely planted. The Middle States do not seem to suit the newer kinds of currants—the Old Red, and White Dutch appearing to do best, and bearing immense crops, properly cultivated, with a very deep soil, rich, rather shaded, and grown in the bush form, which is the most natural and successful—not the tree style. They should be well and properly pruned.

COMMON RED, AND WHITE. Quality, 1.

REMARKS.—Supposed, by some, to be the same as the Old Red Dutch. Very valuable.

KNIGHT'S SWEET RED. Color, light red; quality, 1.

REMARKS.—Varying from Red Dutch, only in the fruit

being less deeply colored, and slightly less acid; resembling more the White Dutch.

KNIGHT'S EARLY RED. Color, red; quality, 3.
REMARKS.—Not very early, as represented to be.

Missouri.
REMARKS.—Only ornamental. Very early, and blossoms very fragrant in Spring.

RED DUTCH. Color, red; quality, 1.
REMARKS.—Not distinct from the variety usually grown in most old gardens. About the very best sort to cultivate in the Middle States, and in our vicinity (Cincinnati).

Striped Fruited. Only as a curiosity.

VICTORIA, *Goliath, etc.* Color, red; quality, 2.
REMARKS.—Bunches, long, somewhat larger than Red Dutch, and slightly more acid; ripens later, and hangs a long time. Plants vigorous, of a spreading habit. Its very large bunches make it desirable for market culture. It requires more trial in the vicinity of Cincinnati. Its lateness in ripening makes it more valuable.

WHITE DUTCH. Color, white; quality, 1.
REMARKS.—Differing only from the quality of the Red Dutch, in being of a yellowish white, and less acid, more delicate in flavor, and therefore preferred for table use. Not quite so hardy as the Red Dutch; rather more delicate in flavor, less acid, and the berries generally larger. Currants are better, in this climate, for a little shade.

WHITE PEARL. Color, pearl; quality, 2.
REMARKS.—A new variety. Bunches very large.

WHITE GRAPE.

REMARKS.—Larger than the White Dutch. Very productive, large, and fine flavored. Makes a superior wine, and when two or three years old it was as good to most palates, as any Port wine.

PALNAU.

REMARKS.—French. Early; productive. Not sufficiently tested.

GOOSEBERRIES.

BRIGHT VENUS. Color, whitish green; surface, hairy; form, obovate; size, 2.

REMARKS.—Flavor, best. Hangs a long time.

BUNKER HILL. Color, yellow; surface, smooth; form, roundish; size, 1.

REMARKS.—Branches, spreading; flavor, very good.

CROWN BOB. Color, red; surface, hairy; form, oblong; size, 1; quality, 1.

REMARKS.—Branches, spreading; early; flavor, best. This is little liable to mildew. It is thought a good plan to plant them on the north side of a board fence.

EARLY SULPHUR. Color, yellow; surface, hairy; form, roundish; size, 2; quality, 1.

REMARKS.—Branches, erect; flavor, best; ripens very early. Nearly all the English gooseberries mildew. They ought to have a cool soil. That is the reason why they grow so well in England. They succeed near Chicago for the same reason; some people near that city have no difficulty with them.

HOUGHTON'S SEEDLING. Color, light yellow; surface, smooth; form, nearly round; size, 2 to 3; quality, 1 to 2.

REMARKS.—The most hardy, productive, and free from mildew, in the country. Fruit, rather small. American. Suited best for cooking. This fruit is much more apt to be free from mildew, when grown in the shade, than in the sun, as the experience of some has demonstrated. The only gooseberry upon which we can entirely depend.

IRONMONGER. Color, red; surface, hairy; form, roundish; size, 3; quality, 1.

REMARKS.—Branches, spreading; flavor, very good. Productive.

PALE RED. Color, pale red; surface, hairy; form, ovate; size, 2; quality, 2.

REMARKS.—Very productive. Flavor, very good. Resembles Houghton's Seedling closely. Like it never mildews.

RED WARRINGTON. Color, red; surface, hairy; form, round oblate; size, 1; quality, 1.

REMARKS.—Branches, drooping.

ROARING LION. Color, red; surface, hairy; form, oblate; size, 1; quality, 1.

REMARKS.—Branches, drooping. Flavor, best. Hangs late.

WHITE HONEY. Color, white; surface, smooth; form, round oblate; size, 2; quality, 1.

REMARKS.—Branches, erect. Flavor, best.

WHITESMITH. Seldom mildews.

APPROPRIATE LOCATION, SOIL, AND TREATMENT OF FRUIT TREES.

THERE is yet a very great deal to be learned on these subjects. This knowledge can be only obtained by practical trials and great attention to the particular requirements, habits, and suitable locality of each kind of fruit. The mechanical and chemical condition of the soil, and its complete drainage, have not met with the consideration due to their great importance and value, especially when the land is too compact in its nature, and will not readily carry off the surplus water. The roots of trees are very apt to be affected in very damp, cold, and hard-pan lands, and diseases will thereby be likely to ensue; deleterious substances being imbibed, and nutriment of a healthy kind can not then be properly elaborated. Too much moisture, and the result, a low temperature, and an imperfect preparation of the soil, will greatly affect the vitality of the plant, and disorder all its functions. In time, it may be gradually *drowned*, if we may be allowed the expression. Numerous derangements, such as the black spots on the fruit, canker, fungous excrescences, and fermentation of the sap, affecting and separating the bark, with numerous other evils, may, probably, arise from this cause. Subsoiling with the plough, and trenching with the spade, should go hand in hand with the draining materials, and open the ground to the salutary influences of the air and light to destroy the injurious acidities of the earth, elevate its low temperature, and render available its unexcited riches. It has been stated by a writer of high authority, that he frequently found the soil of a *well-drained* field higher by the thermometer, from ten to fifteen degrees, than that of another field, not so drained, though in every other respect the soils were similar. Draining is also the very best preservative against the drought. With regard to situation, there are some kinds of fruit trees that, like cats, are very strictly local in their habits; while others are more ubiquitous, and a minority of them are at home everywhere. A suitable stock is also necessary for every graft to arrive at the desired normal condition. It has been the paramount object of the writer of this work, and the principal reason of its being undertaken, to make the description of the fruits therein contained, subservient to the wants of each sort; and the particular attention bestowed in bringing forward the reports from all quarters, but more particularly from the Western States, and our own neighborhood, will, we think, sufficiently prove it. With the careful, though generally brief, portraiture, individually, of our chief pomological treasures, there will be seen the particular locality to which each is specifically adapted, thus rendering the path of the cultivator more clearly defined, to enable him to discover what he should obtain, and what he ought to avoid. The action of Pomological Societies, but more especially that of the Cincinnati Horticultural Society, has been of inestimable advantage to us, especialy from their lists of "worthy" fruits, of which it has been seen we have so largely availed ourselves, and so liberally presented to our readers. And the labor and money saved to both nurserymen and fruit-growers by the Cincinnati Horticultural Society alone, in its reports of those varieties which are "unworthy," or rejected, are also incalculable.

And now a word in relation to "Dwarf Pears." We presume that although it has not yet been most conclusively and satisfactorily proved (owing to the time not having been had) that our Western climate and soil is as favorable to the pear on the quince, to say nothing of the pear on its own stock, as the Eastern portions of our country—Massachusetts for instance—yet we think we may venture to say that there is but little probability that the difference is so great, in this regard, as to act as a barrier to their successful cultivation, when we shall have taken the pains to avail ourselves of all the requisites for that object, in every other respect. Col. Wilder, as much of a monarch among fruit-raisers as he is a prince among merchants, has given us ample proof in his own case, and that of many others, that they have reached success in this department of pomological science, in his communications to our Society, and to others, on this subject. We have no reason to doubt his statement, corroborated, also, by eye-witnesses from among ourselves. He suggests "that the *pear, upon the quince*, should be planted deep enough to cover the place of junction, three or four inches below the soil, and then the pear will throw out roots from itself, and the result will not only be an early fruiting, but also longevity; and (the Col. adds), to obtain the pleasure and profit of regular crops, for many years, before the trees would produce fruit on their own stock." We believe this all right, and an advantage; the only objection that occurs to us is, that as the trees, when they take root from their own stock, cease in a great measure to be dwarf, they will take up more space in a garden, etc., than they would if confined only to the quince stock, and so far, in a measure, defeat the object intended.—Figures, like facts, are stubborn things. They have made out the profits of these trees in the East. Our good time may come yet, when, notwithstanding the unfavorable balance against us, if any, in climate, we shall have complied with all the conditions necessary for their prosperous culture. The celebrated Mr Beeckman, formerly of Belgium, but now of New Jersey, says that quince-grafted pears are less subject to blight, that pernicious pest which has destroyed the hopes and prospects of so many. He says:

1st. "Have a good, substantial, rather deep soil, with porous or drained sub-soil.

2nd. "Select the Angers or Orleans Quince (Paris is good with us in the West, and probably the strongest).

3rd. "Plant no other varieties than those which succeed on the quince.

4th. "Plant the trees deep enough, so that the place where they have been budded shall be at least three inches below the surface of the soil.

5th. "Keep the weeds down.

6th. "Keep the branches low, and make a pyramidal tree, by judicious pruning once or twice a year. If well pruned, the tree requires no 'pinching.'"

We shall conclude with a few observations on the selection of trees suitable for an apple orchard, which we have drawn in a condensed form from a series of papers on the subject, from the pen of Dr. Jno. A. Warder, President of the Horticultural Society; an excellent writer, who has improved his natural gifts and tastes by much study and observation, and who has long devoted his talents to one of the noblest pursuits that can occupy the mind of man. The articles from which we extract were published in that capital and too neglected work, the *Western Horticultural Review and Botanical Magazine.* The general subject is the selection and planting of an apple orchard. In planting an orchard for family use, he says:

It is a very common mistake to plant too many varieties. We are apt to select generously, rather than judiciously; few men are capable of making out a select list of trees that shall bear a succession of fruits for the table and kitchen, so as to have a constant supply during the year—such a selection requires a pretty thorough knowledge of Pomology, and great familiarity with the varieties, and their adaptation to different soils and situations.

For a SINGLE FAMILY, a dozen trees should furnish a superabundance of fruit; but the observation of all who have attempted a selection is, that the smaller the list to be chosen, the greater will be found the difficulty in making the selection. The following list is recommended, commencing with the earliest: the Red Juneating, Prince's Harvest, Summer Rose, Fall Pippin, Rambo, American Golden Russet, Newtown Spitzenburg, White Bellflower, Swaar, Pryor's Red, Raule's Janet, Newtown Pippin.

Others would prefer, White June, Benoni, Strawberry, Golden Sweet, Fall Pippin, Rambo, Westfield Seek-no-further, Newtown Spitzenburg, Yellow Bellflower, Waxen, White Pippin, Roxbury Russet.

FOR MARKET, the question assumes another phase altogether, for it is a matter of great importance to furnish a supply that shall suit the tastes of customers, and the consumers are possessed of as much diversity in this respect as in any other; and withal, are coming rapidly within the influence of the enlightening rays of the Pomological Conventions of the States and Nation; although still bound by fashion and early association. Thus, in New England, the Porter, Baldwin, Rhode Island Greenings, Westfield Seek-no-further, and Boston Russet, will outsell most other varieties. In New York, the Greening, Esopus Spitzenburg, and the Newtown Spitzenburg (their Vandevere), the Newtown Pippin, and the Yellow Bellflower, will bear the palm; and in Philadelphia, another style of fruit would command more attention, among which the Rambo would stand pre-eminent; and the beautiful little Lady Apple, which constitutes a necessary part of the Christmas decorations, and the ornaments of every winter feast, will receive the highest price of any other Apple in either of those cities.

In the SOUTHERN MARKETS—to which vast quantities of the products of our Western orchards are shipped by river—all red apples are preferred to those of lighter color. The warmth of the climate also requires that the fruit should be of the firmest and best-keeping varieties; hence we find that the Pryor's Red, the Raule's Janet, and even the Gilpin, are favorites among the shippers and purchasers, who will almost at once reject all of the light-colored fruits.

Near our large towns and cities, particularly in the vicinity of railway stations, a very large supply of choice fruit is wanted for the retail trade throughout the season, and the retailers desire fruit that shall be attractive, and pay much less regard to real excellence than to external beauty—so that it may be profitable to produce a supply for this demand, that may not rank as first-rate, but realize to the producer the highest prices. Thus, for this purpose, the beautiful Maiden's Blush, which can scarcely be ranked as *good*, will be purchased readily, while other and better sorts remain unsold. The Waxen will follow after this fruit, and be eagerly sought on account of its appearance; the Yellow Bellflower always commands a high price, though not admired universally as a dessert fruit.

The HOTEL KEEPER, in purchasing, has two objects in view; the decoration, as well as the supply, of his table; and he cares less about the quality than the appearance of the fruit he purchases. For a dinner of one hundred plates, it is cheaper to pay one dollar for a bushel of neat, medium-sized, or small apples, containing one hundred specimens, than to give one dollar and a half for two bushels of the finest and largest fruit of the season, that shall count but one apiece for the party. This matter of calculation is beginning to be well understood, and even private families find it to their account to select medium-sized fruit, independent of the fact, that most of the apples of the highest character are not large. The best dessert fruits are of medium size, although many of the large kinds are good for the table, and may be much preferred for the kitchen.

From these hints, it will be inferred that it is important, before setting out an orchard for profit, to determine the character and extent of the demand to be supplied; and also, whether it be northern, eastern, or southern, or even beyond the seas, since each will require a different selection of varieties. In all cases, it is safest to select *very few varieties*, and those the varieties that have proved successful in your particular neighborhood—for it has long since been observed, that the finest fruit of one region may become quite an inferior affair in another; nor can any one determine this point, beyond a general guess, without absolute observation or trial. A change from a somewhat sterile soil to one of great fertility, is generally followed by an increase of size, but often, also, by a greater coarseness, in appearance at least. So remarkable is this, that pomologists are often puzzled in recognizing fruits produced in our region, although they may be perfectly familiar with the same variety at home.

There is one variety, so far as the trade has yet been tested, which will pay well for transportation to the English market; it is known there as *The American Apple*, and here as the Newtown Pippin. Mr. Pell, and other orchardists, have found it the best variety for shipping on a long voyage, and others may safely rely upon their experience; but in some sections the Newtown Pippin does not succeed so as to be profitable. As a general rule, it is claimed that this tree should have a rich calcareous loam, but the subsoil should not be too tenacious and wet.

The growing of apples as a food for stock has, within a few years, attracted much attention among the most intelligent agriculturists; and pomologists have been engaged in selecting lists of such as were most hardy, productive, and ripening in succession. Those who have tried the most experiments in the use of this kind of food, speak in the most exalted terms of the results.

The following sorts of sweet apples are recommended for a succession of fruit for stock feeding; they are particularly selected for their hardiness and productiveness, and are offered with the restrictions and provisos already noted: Sweet June, Sweet Bough, Golden Sweet, Jersey Sweet, Baily Sweet, Ramsdell, Danver's Winter, Talman's Sweet, Michael Henry, and Campfield.

For a Cider Orchard.—There are several varieties that have been fully proved to be superior to most others. They should be of thrifty growth, very productive, and should ripen late in the season. It is a great mistake to suppose that any common apples will answer for the manufacture of cider.

The varieties that have been found most desirable for this purpose are those which contain the most saccharine principle, and which will yield the heaviest must; they are not, however, necessarily sweet apples—such are the Harrison, Campfield, Graniwinkle, Gilpin, and especially the Hewes' Virginia Crab. Several others, on account of their hardiness and productiveness, are frequently planted for cider orchards.

Horticultural Society's selections for GENERAL PURPOSES:

For July and August,	2	Summer Rose.
" July and August,	2	Strawberry.
" September and October,	2	Fall Pippin.
" October and November,	4	Rambo.
" November and December,	5	Golden Russet.
" November and December,	5	Newtown Spitzenburg.
" November and December,	20	White Bellflower.
" January and February,	15	Pryor's Red.
" February to April,	40	Raule's Janet.
" March,	5	Newtown Pippin.
	100 trees.	

Having selected the trees, the next points to be considered, are,—When, and how to plant them? Two very important questions, upon the answers to which much of the success of the plantation inevitably depends. As to situation, it may be said, generally, that whenever it is possible, select an elevated position for the orchard; though no one should be discouraged, or deterred from planting, even though he be located in the level bottom lands of our rivers. Those who would establish large orchards, with a view to profit from the sales of fruit, should, by all means, be advised to select their sites in elevated and hilly regions, on account of greater immunity from frosts, the finer fruits that are produced with the better circulation of the air, and a soil generally better adapted to the purpose of fruit-growing.

One special advantage of elevated positions for the orchard, consists in the character of the soils that are generally found in such situations—they are less rankly fertile, and less productive of excessive wood-growths, than the deep alluvials that prevail in the bottoms. This appears to be a condition most favorable to the highest development of the apple in its greatest beauty and perfection. Upon hills there is generally a more probable immunity from the lichens, that so often disfigure the fruits produced in the damper atmosphere of lower levels, which is also less stirred by the winds. The very texture of the fruit, and its consequent keeping qualities, are undoubtedly superior upon the thinner soils of the hill-lands, than elsewhere. These are mere matters of observation, now cumulating and not to be gainsayed, but rather spread out before the public, for the sake of guarding them against mistakes in selecting the sites of large commercial orchards, than with any desire to prejudice the thousands in the country which are planted in the other class of sites and soils, for the very good reason that they must have been planted there, or not planted at all. There are soils and situations, however, which are so bare, broken, and rocky, as to be unfit for cultivation; if such are planted as orchards, it will be unreasonable to expect so heavy, or so fine crops of fruit as where judicious culture can be applied.

Having settled the question of situation, that of soil ensues as a matter of considerable importance; for no one can doubt that the permanency and success of the orchard will depend very much upon whether the trees find sufficient and proper aliment from which to elaborate their desired products. In this particular, however, we are relieved from any great anxiety, by observing that the apple-tree thrives in almost every soil that contains the usual mixture of materials that is common to all reasonably fertile land: certain definite elements are, however, necessary; among these are potash, and phosphate of lime, which enter largely into the constitution of the plant and its products, and should exist in the soil, or they must be applied artificially.

Too little attention is paid to the preparation of the land; and it is believed, and has been fully demonstrated, that a thorough tillage before planting is very conducive to success. And for three or four years after planting, the ground should be kept in tillage, either with or without root-crops, avoiding the introduction of cereals, for the double purpose of preventing the consequent abstraction of elements that will be needed by the future crops of apples, and also, that the constant culture of the soil by tillage may not be interrupted.

After the second or third year of tillage-crops among the young trees, which will have encouraged them to make a vigorous growth, some orchardists will find it advantageous to sow a green crop, such as buckwheat, oats, or peas, to be pastured off by hogs, before they ripen.

It is believed that the green manure, thus left with the animal droppings, will be very advantageous, and may be plowed in with good effect. The swine will not be apt to injure the trees materially, unless they are kept on too long, so as to pasture the green crop very closely. In all the tillage, the greatest care must be exercised to avoid wounding the bark by carelessly striking the stems with the single-trees, when the plowing is done with horses; an additional hand will be needed in plowing next the trees, to save them. When the trees have made a good growth, and bear well, the tillage may be suspended, and the ground sown to clover, to be pastured with swine for a year or two, and again broken up and tilled.

Many orchards have been planted in grass lands, however, without any especial preparation of the soil, as above recommended. When this is necessary, there should be a substitute for the thorough and continuous tillage. The ground about the trees must be turned over with the spade, in a sort of rough-digging, twice or three times a year; in the Fall, or early Winter, in the latter part of Spring, and again during the Summer, so as to make the earth loose and mellow, and to keep down the weeds. This digging should be about six feet in diameter, and gradually increased, as the roots extend; it should be shallow next the tree. In some situations this kind of treatment will be necessary, and must be adopted as the best that can be employed; but thorough and continuous tillage with the plow and cultivator is recommended for several years; indeed, some persons insist upon constant culture of orchards, as conducive to their best success; one or two plowings, during the season, with a naked fallow, or green crops turned in.

The distance at which the trees should be placed, will depend upon the soil, and also, in a great degree, upon the character and habit of the variety; since some kinds are much more thrifty, and grow to a larger size than others. As a general rule, the richer the soil, the greater the space that will be required. From this suggestion a valuable inference may be drawn:—that trees, particularly for large orchards, should be studied, in regard to their style of growth, and assorted accordingly, before planting. Thus, there are some kinds that will do better if set at twenty-five feet apart, than others planted at forty-five feet distance. As a general proposition, trees should not be too much crowded; and when the price of land is of small moment, compared to the success of the orchard, forty feet may be assumed as a good average distance—two rods is a very common allotment.

Below is a list of some of the coarsest growing, compact, and least diffusely branched varieties:—

For wide planting—Rhode Island Greening, Pennock, Summer Queen, Newtown Spitzenburg, Vandevere Pippin, Roxbury Russet, Fall Pippin, Yellow Bellflower, Fallawalder.

For close planting—American Golden Russet, Juneating, Lady, Raule's Janet.

Laying out the ground, will afford an opportunity for the exercise of some taste and fancy, and will require accuracy. The usual form is that of a square, or in rows, crossing one another, at right angles. But many advise wide planting in this manner, and the addition of a central tree between each rectangle; this is called the *quincunx* method. The hexagon style has also been suggested, but the rectangle affords the greatest convenience. Whatever plan be adopted, the places or stations for the trees should be marked by a stake, which will indicate the positions where the holes are to be dug.

The stations should be prepared by opening large holes, if the ground has not been thoroughly prepared by deep plowing—if, however, the soil be mellow, the excavation need not be larger than the roots of the young trees require for their accommodation; from eighteen inches to two feet square, and about one spade deep. If, however, the land be in grass at the time of planting, the holes should be made four or five feet in diameter, and should be deeply excavated.

Planting the trees, will require the exercise of great care and judgment. If the stations have been well laid out, and carefully excavated, the trees may readily be made to range with that perfect regularity which gives a pleasant character of artistic propriety, that will be a source of satisfaction for many years. To insure this result, it will be best to set range stakes in either direction; after a few trees are planted, they will aid in the correct setting of the remainder.

We shall not need the theories of the terraculturist, but simple observation, to induce us to plant the trees at the same depth they previously occupied in the soil—the *collar* should be at the surface. To this end, the hole should be partially filled with good mellow soil, a little raised in the middle, and upon this bed the tree is to be placed, in its proper position in regard to range and depth. Then the finest mold is thrown lightly on the roots, after they have been carefully spread out. Gentle pressure by the foot, will aid in setting the earth about the roots, after taking particular care that the fine earth has been well worked in among the fibers. Some planters recommend the application of a bucketful of water at this stage, particularly if the soil be dry; but others consider this of doubtful propriety, particularly when the land is stiff and clayey. When the natural soil is poor, or unkind, and not well prepared, it will be necessary to use a good compost for filling up the holes; but great care should be taken to avoid the application of any strong or fresh manure. Rotten sods, with old decayed chip manure, or cow-yard scrapings, well incorporated, and treated with lime and ashes, or some other judiciously-prepared compost, will be available, but should have been prepared beforehand. In filling up around the tree, less care will be requisite as to the character of the earth. The surface should be made a little rounding, to allow of settling, and also to prevent the accumulation of water in a wet season; this is to be particularly observed in Fall planting. Carefully examine the roots, and remove, with a sharp knife, all those portions that have been unfortunately torn, or wounded by carelessness, in digging at the nursery, or in transportation. Set the roots upon a bank of mellow earth, spread out every fiber in its natural direction, fill in with the most mellow soil, or compost, shaking the tree very gently, and working in the dirt thoroughly; fix the roots by a gentle pressure of the foot, and then fill up the hole to the proper level of both tree and surface. In dry weather it may be necessary to leave a slight concavity or hollow about the stem, so as to retain moisture until it can soak into the earth; but in newly-planted trees it is better to sprinkle the tops. *Mulching*, or covering the whole surface above the roots with straw, etc., to prevent the evaporation of moisture, will be found of inestimable value, particularly in a season of drought.

Subsequent Summer treatment will depend, in a great degree, upon the manner in which the ground had been prepared for the young orchard. According to the best authorities, the soil should have been thoroughly loosened by deep plowing, and, if practicable, subsoil plowing, in most soils, before the trees were set, for such complete culture can never afterward be applied, on account of the presence of the trees. If this kind of preparation has been made, it will be best to continue stirring the whole of the ground, with the plow or cultivator, so as to destroy the weeds and maintain a good tilth during the season;

this will require two or more repetitions of the plowing, according to the character of the soil, and the abundance of weeds and grass.

Cultivation of the soil among the trees should, if possible, be continued for some years, to secure and promote the rapid and healthy growth of the orchard, even should it be objected that such a thrifty growth of the trees is not followed by early productiveness. Lay a broad foundation for future bearing of large crops, and the coming years will not have to blame you for stinted trees, unable to produce a liberal yield. The plow is the great agent of culture, upon which we depend for the comminution of the soil, and its subversion, when we desire to bury the weeds and expose the earth to the influence of the atmosphere, which is ever ready to impart its aerial treasures of gaseous manures, from which, indeed, the chief element of woody fiber is to be derived. Two or more plowings of the soil will be found necessary, and will maintain the requisite mellow condition and freedom from weeds.

In all cases, the use of the plow and cultivator, especially the former, should be guarded with great care, to prevent the injurious contact of the trace-chains and whiffle-trees with the bark, that would be otherwise bruised and often removed, to the great injury of the growing trees. The single-trees should be as short as possible, and as the team approaches the rows, an assistant should watch and guard the young trees, by lifting up the projecting portion. Some persons prefer oxen for this culture among trees, upon the supposition and belief that they are more readily controlled in their steady gait, slower than horses, and because the yoke alone is liable to injure the trees, and this is more perfectly under control of the driver. For heavy plowing, the oxen would be preferred by most operators; but for the light, continuous culture among young trees, the horse, or, perhaps, still better, the small-footed and precise-stepping mule would be preferred.

When plowing among orchards, care should be taken not to open a land between two rows repeatedly, by throwing the first furrow against the tree, but the lands should be alternately opened and gathered, so as to maintain a level surface; unless, indeed, the surface be very flat and humid, when it may have been necessary to plant the trees upon the original surface, and cover the roots with a small hillock of earth; here, the constant "opening" of the lands, by throwing the furrows toward the trees, will be advisable; this soon supplies a series of superficial drains that are beneficial to the orchard.

Cultivation of the soil may be continued for several years, with the best results, as will be very apparent in the thrifty growth, fine foliage, and smooth bark of the trees; but it is very desirable to have it continued at least during three or four Summers; after which the surface may be laid down to grass, provided due care be taken to keep a large space cleared about the trees, but no cattle should be allowed to pasture upon the land, except swine, which will destroy insects that are in the fallen fruit.

In some situations, whether from convenience or necessity, the young trees are set in a grass field, which may be rocky, or otherwise unfit for the use of the plow and cultivator. When this is the case, as general culture of the whole area is impracticable, it becomes advisable to pay special attention to the treatment of that immediately occupied by the roots. When planting in such a field, the holes should be dug much larger than required to receive the roots, and, of course, much wider than when planting a thoroughly prepared soil. The grass and weeds must be kept under control by digging about the trees to an extent of five or six feet in diameter, or more. This digging may be performed during the Winter, when the frost will permit, and will require repetition by Midsummer, and perhaps again during the season, unless we have applied a most excellent adjuvant, the mulching of the surface.

So much has been said and written of late years upon the subject of *mulching*, that it must be familiar to all, yet a few words will be in place. The object of mulching is to preserve a certain degree of moisture in the soil, about the roots, by preventing the rapid evaporation from the surface, in our arid climate; it will, therefore, be particularly serviceable in sandy and gravely soils. The mulch may consist of any light rubbish that may be at command; straw is generally used, and is applied to freshly-dug soil to the depth of three or four inches; this soon settles down and forms a close coat that prevents rapid evaporation, collects dew, and with it, ammonia, which it retains, ready to be washed down into the soil by the next rain. Chip dirt, from the wood-pile, or old tanbark, are very suitable materials to be thus applied, and even loose brush and twigs, the trimmings of the trees themselves, which will retain the blowing leaves, will form a good mulch, and keep the soil loose and mellow beneath; but where accessible, as from a saw-mill in the neighborhood, there is nothing that so well and so neatly produces the *desired effect as coarse saw-dust, which, in stiff soils, may* be turned in at the winter digging, or scraped to one side, and reserved for application the ensuing season.

Young trees, freshly set out, especially those from crowded nurseries, and where they have been cleanly trimmed up with naked stems, just such as are generally most admired, on account of their resemblance to walking sticks, perhaps, are frequently obnoxious to serious evils when transplanted to open exposures. The smooth bark is often scorched and blistered, and oftentimes fine young trees are also destroyed by the larva of an insect that lives upon the cambium or young wood just beneath the bark; the worm eats away a considerable surface, often nearly girdling the tree before his invasion is discovered. The latter injury can only be suitably met with the sharp knife, in the hands of the ever-wakeful orchardist, who watches his trees with lynx-eyed vigilance; but the former evil may be prevented by a very simple contrivance, and one that is particularly recommended for the cherry, that has its bare and polished shank exposed to a hot sun, after having been drawn upward in search of light and air in the close rows where it grew. The application consists in a wisp of straw gently tied to the stem, and extending from the branches to the ground. As the trees grow larger, the straw may still be applied with advantage to the cherry, which appears to suffer from sunshine, both in the Winter and Summer, more than other trees; for the larger trees, a straw rope is used, wound about the stem; but a couple of boards tacked together, and set up to the south and south-west sides, have been found a very efficacious protection.

Low-headed trees are, on many accounts, to be preferred in our climate, even for the apple, the great orchard fruit. These should have their training commenced in the nursery, but it is seldom there attempted, on account of the desire with most purchasers to see tall trees: often mere whip-stalks, trimmed up clean and straight, will sell more readily than stout, stocky young trees, containing every element of future beauty and usefulness. Always select such when it be possible, remembering that we have already agreed that the ground planted in fruit should not be appropriated to pasturage, and hence the tall stems are not needed to keep the fruit and foliage up out of the reach of cattle.

Having selected properly grown trees and planted and tended them as already advised, the Summer pruning for the first few years becomes a matter of great importance.

A frequent examination of their condition should be made during the growing season, and with good judgment and small sacrifice of wood, great good may be effected. This should consist in stopping rambling or rampant shoots, either by pinching their buds with the thumb and finger, or cutting them back with the knife; here, however, is the point to exercise great judgment. In branching the tree it should be an object, from the first, to divide the head among more than two main limbs, since the division into only two is more apt to be followed by injury from splitting in after years, from the weight of the fruit and foliage, than when the strain is more divided.

SELECTION OF TREES FROM THE NURSERY.

E. J. Hooper,—Dear Sir:—At your request, I sit down this evening to address a few remarks to the readers of your valuable work on fruits, upon the selection of trees from the nursery.

At different periods, and in different places, I have addressed nurserymen upon a similar topic; not exactly the same, but similar, that is, upon the proper mode of growiug trees for sale. These addresses have been received, with different degrees of favor and disfavor; the nurserymen sometimes admitting the truth of the remarks offered, but asserting that purchasers desired to buy their trees by the *foot* in *height*, and not by the *inch* in *diameter*. This being too much the case with planters, as you are very well aware, the attempt is now made to reach them, the buyers; as we may be well assured, that the intelligent nurserymen of our country, very well knowing the true philosophy of the matter, will gladly supply the public with a better article, if that public can be made sufficiently intelligent to demand well-grown instead of tall-grown trees—substantial stocky plants, instead of slender drawn, feeble, whip-stocks.

Those who are about planting an orchard, will do well to visit the nurseries, and see the stocks; then they can select for themselves such trees as they may prefer. Allow me to suggest to them a few things by the way, and before they become fascinated with the tall, smooth stems of the saplings they may be about to visit, and before they may have selected their stocks; and also allow me to send a telegraphic dispatch to the nurseryman, to the effect that intelligent customers are on their way to the nurseries. To this effect, therefore, I will say to the visitors, that it is presumed they, as intelligent planters, and promising orchardists, desire to form orchards, with low heads, having appropriated a certain piece of ground to the production of choice fruit, and not to a general range for the pasturage of all sorts of stock; for no sensible man will expect to produce good fruit without due care of his trees, and the entire devotion of the soil to the orchard.

There are those who can not afford to relinquish any ground to trees alone, they are patriarchal in their habits; they keep droves of cattle which range everywhere, and they have no idea that an orchard field should be inclosed perfectly; and that from it, not only foraging boys, but feeding cattle also, must be excluded. Such persons require that their trees, when planted, should be already fully grown, and that they should be high-limbed, so as to be without branches, until above the reach of all hungry cattle; to such, it behooves us to be lenient, but, at the same time, let us hope that the number of such planters will become less and less from year to year.

Having reached the nursery, with a list of varieties wanted, already made out, proceed to examine the stock, to see if any have been properly grown for the making of an orchard of low-headed trees. The nurseryman, who has been expecting you, will have a *corps de reserve* to suit your case, even though his anxiety to serve the *majority* of customers, and perhaps to gratify his own taste for *tallness*, may have induced him to have a majority of his trees made after the whip-stock fashion. Look at these better trees, and among them proceed to select thrifty, stocky *plants*, furnished with twiggy branches from near the ground. It is not necessary that the side branches should be large, much less that they should be at all equal, or nearly equal to the leader, which should always be supreme, among the branches. Here you may be able to find the desideratum for which you seek. All well-grown trees, to be well-grown, must be developed on all sides alike. With our modern views of the importance of the doctrine of the individuality of buds, it becomes necessary so to arrange the trimming of young trees as to provide an equal and universal, or general supply, over the extent of the infant tree, if we desire to have it well and fully developed. We all know that where young trees are crowded closely, or where they are cleanly trimmed up and crowded together, they will necessarily be tall, slender, and poorly developed, while, if they have had room to develop themselves fully, and have been encouraged to put out lateral branches, they will, perhaps, be less tall, but they will be more stout and stocky; and experience shows that such trees will be more able to withstand the shock of transplanting, and will be much more likely to grow well, just in proportion to the number of vital centers they may possess; these centers are the buds, and the more widely they are distributed over the plant, the better will they be able to exercise their functions. This is not all—that most deceitful and insidious of all the enemies of a young orchard, the bark worm, seldom attacks any trees but such as have a clean, smooth, uninterrupted bark, open and exposed to their ravages, which, alas, are never to be observed until too late, when their work of destruction is completed, and the dead bark, sloughing off from the wood, presents itself to the eye of the orchardist, at the end of the season, and explains, in terms of unmistakable plainness, the damage to which the unsuspecting planter has been subjected. This enemy, rarely, if ever, dares intrude upon a tree that has been properly grown, and which is furnished with lateral branches, from near the root to the proper head of the tree; and if, perchance, it have ventured to intrude, the numerous points of vitality near at hand are able, by their resiliency, to restore the loss, and to repair the damage.

Heretofore these expressions have applied to the generality of trees; the bark worm, however, is peculiarly destructive of the apple, quince, thorn, and pear, perhaps in the order named. There are, however, other evils that are inflicted upon high-grown trees, that are still more apparent with the cherry and other sorts of fruit. The effects of the bright sunshine upon such naked stems is also disastrous, as seen in the dried, and shriveled and scorched bark of many a noble, tall, but naked specimen—while those but partially shaded by the moderate spray that exists in well-grown trees, entirely escape these results. It is a common remark with regard to the cherry, particularly the freeer-growing sorts, that they will suffer from bark-bursting, which is attributed by some to the frost, and by others to the influence of the sun. Whoever saw these effects, from what cause soever, upon low-headed trees, or above the branches? No one often, though there may be exceptional cases. No, these results are always most manifest upon the prettiest, most thrifty, and most *naked* trees; and the very remedy which has proved most effectual is simply supplying to such trees as are most exposed, the shade by boards, or straw-wrapping, that might have been afforded more cheaply and better, by the natural limbs of the tree. In selecting cherry trees, therefore, choose those which have been so treated, in the nursery, by heading down, or otherwise, as to have their sides furnished with low branches, instead of those which have been either trimmed up high, or allowed to branch only from the upper buds; so as to make a high head. With pears, apples, and most other fruits, the same rule will apply, and you may safely restrict your selection to those that are furnished with moderate side branches from near the ground. Rest assured, that it will be much more easy a matter to trim these lateral branches off, as you may wish to raise the head of the tree, than it will be to produce them from one of your beautiful, smooth, high-trimmed trees, such as are constantly to be seen, and which, unfortunately, are so much admired.

With the peach there is less importance to be attached to this matter of nursery trimming; for the virgin trees are always of but one year's growth from the bud, and it matters little how they may have been branched and have grown, for a judicious planter will always cut them down to a foot or less, when setting them out, and force them to take a new start. The more equable this is, the better for the future tree; for, of all our fruits, the peach most needs to be a bush, branching equally from its stock, near the ground; but this plan involves the modern admirable ideas, of pinching in, and shortening of fruit trees, particularly applicable to the peach, and this is a topic that it was not intended to broach.

These general rules with regard to the branching of the trees being premised, next look to the condition of the bark, which should be smooth and fresh, not wrinkled, nor mossy with age; for all thrifty young trees will have a healthy, smooth skin; this part being to them, as to animals, an organ of great importance in the function of transpiration, and here, too, in that of respiration.

When taken up from the ground with sufficient care, the roots should present an appearance of a mass of fibers, rather than that of a few prongs of smooth forks. The former condition is that most favorable for success in planting, and is worth much to the purchaser, but is not obtained without the expense of labor and time by the nurseryman, and should always entitle him to your confidence, and to a liberal remuneration, for the extra pains he has taken in transplanting the trees to produce this condition.

Excuse, dear sir, these hasty notes, thrown off rather as suggestions than as precise rules for the selection of trees. Hoping that they may not be entirely useless to some of your readers, believe me yours,

Feburary 23, 1857. JNO. A. WARDER.

PRESERVATION OF FRUITS, BY SCHOOLEY'S PATENTED PROCESS.

As we have referred to Schooley's Patent Process for the Preservation of Fruits, we deem it our duty to say a few words in reference to it. We had the pleasure and satisfaction of examining the plans of Mr. John C. Schooley, of Cincinnati, several years ago, and were then confident they were based upon correct scientific principles. Since our first examination of his plans, Mr. Schooley (who has endeavored for many years to ascertain the best mode of preserving ripe native and tropical fruits) has been very successful, and has demonstrated, that the ripening process can be so retarded as to preserve all kinds of our choicest native fruits, from one season to another, with all their original flavor and freshness. It is evident that various means have been used during the last fifty years, whereby ripe fruits could be preserved from year to year. Ice has been used in many ways, which, in every case, succeeded in reducing the atmosphere to a proper temperature; but the great difficulty has been, to get clear of the moisture, and then again to produce a continued supply of desicated, or dry, cold atmosphere.

If fruits are placed in a room containing moist, still air, though it be sufficiently cold, these fruits will become moldy and musty, and if retained any length of time in this atmosphere, decomposition will inevitably take place, and it is evident that this air will become contaminated with such substances as the aqueous vapor holds in solution; and when there is no current of fresh air, all substances submerged in such an atmosphere will become enveloped in their own exhalations. Hence, a current of pure, cold, dry air is essential for the preservation of all organic substances, and particularly ripe fruits. In a fruit-room, constructed upon a large scale, Mr. S. has been enabled to preserve quantities of our most tender fruits, during the space of one year, with all their original flavor. The entire process, as patented and put in operation by Mr. Schooley, is fully indorsed by the American Pomological Society, as will be observed by their last Annual Report; and the president, Hon. Marshall P. Wilder, in noticing this valuable invention, in his annual address before the society, says:

"Having heard of the great success of Mr. Schooley, of Cincinnati, Ohio, in his celebrated discovery for the preservation of meats, I opened a correspondence with him with respect to the application of the same process to the preservation of fruit. He subsequently visited me at Boston, and advised as to the construction of a fruit room upon his principle. This I have found, during the last Winter and the present Summer, to operate in accordance with his statement, as illustrated by Prof. Locke, in his 'Monograph upon the Preservation of Organic Substances. By his plans, the temperature and moisture of the fruit-room, and consequently the ripening of the fruit, may be perfectly controlled. One gentleman informs me that he kept strawberries in a fruit room constructed on this plan, from June 1st to the 20th, in perfect condition for the table; and he entertains no doubt of its complete success in the preservation of apples and pears indefinitely. My own experience corresponds with this statement."

We do not hesitate to recommend Mr. Schooley's patent to the attention of all fruit-growers, as an invention of great importance; and properly constructed (according to the plates published in the first part of this work, and taken from the Annual Report of the American Pomological Society), can not fail to be successful.

THE END.

SCHOOLEY'S

CELEBRATED

MEAT, PROVISION,

AND

FRUIT PRESERVER.

PATENTED MARCH 13, 1855.

A rare opportunity is offered to any one desiring to purchase the entire Right of my Patent for any one State or District. I ask no one to embark in the enterprise, unless he has most thoroughly examined its value and usefulness. I do not hesitate to say, that it is one of the very best Patent-rights ever offered to the public. Its success is unequaled, as will readily be observed by the hosts of recommendations and testimonials now in my possession; some of which can be seen in Prof. Locke's Monograph (a copy of which will be forwarded to any address, on receipt of eight three-cent postage stamps). I also refer to the Annual Address of the Hon. Marshall P. Wilder, of Massachusetts, before the United States Pomological Society, held at Rochester, N. Y., in September, 1856; also, to the published Annual Proceedings of the same Society, of that year.

I have received *Letters Patent for England, Ireland, Scotland, and the Channel Islands; also for the Empire of France;* all of which Patents I offer for sale, with the exception of the Refrigerator Right for the Western, and a part of the Middle and Southern States.

Every Farmer should have one of my Fruit-Houses attached to his Orchard. It can be done with little or no expense, and will be a saving of hundreds of dollars per annum.

Individual Rights to Farmers or Fruit-Raisers can be had on very reasonable terms.

Exclusive Rights of Counties would be an object to any Fruitier, as he could then control the business for that region.

Parties desiring a single Right, will be particular in giving precise location (side-hill or level ground), dimensions, quantity of ice to be used, etc. I will sell the exclusive Right of one county for a small advance over the price of a single Right.

INFRINGEMENT.

In the Annual Proceedings of the American Pomological Society for 1856, will be found a full description of my Patent Fruit-Preserver with elaborate drawings of same. The compiler of that volume unintentionally omitted to mention to the public that it was patented, and for fear many may be misled and feel themselves at liberty to build and use the same without my permission, notice is hereby given to all parties who will either make, sell, or use any mechanical structure, operating in the manner as claimed by me, without permission from me or my assigns, that all such will be prosecuted at once, to the extent of the law.

For the purchase of Rights, and all particulars in regard to the Patent, address

JOHN C. SCHOOLEY, *Patentee*,
Cincinnati, O.

AGENTS WANTED.

I wish to employ several energetic Agents to travel through the Eastern, Western, and Southern States, to sell Fruit and Pork-House Rights. Address as above.

www.ingramcontent.com/pod-product-compliance
Lightning Source LLC
LaVergne TN
LVHW010207110826
845151LV00002B/626

* 9 7 8 1 4 2 5 5 3 5 6 6 7 *